SOVIET
TRANSPORT AIRCRAFT
SINCE 1945

The Polar Aviation Il-14 SSSR-04179 at the SP-12 (North Pole 12) station on the drifting Polar ice, in 1964.

SOVIET TRANSPORT AIRCRAFT
SINCE 1945

JOHN STROUD

FUNK & WAGNALLS
New York

© 1968 John Stroud
All rights reserved
Library of Congress Catalog Card Number: 68–21860
Published by Funk & Wagnalls
A Division of Reader's Digest Books, Inc.,
by arrangement with Putnam & Co., Ltd.
Printed in England

CONTENTS

Introduction	vii
Acknowledgements	ix
Soviet Air Transport	11
Antonov An-2	41
Antonov An-8, An-10 and An-12	50
Antonov An-14 Pchelka (Bee)	65
Antonov An-22	71
Antonov An-24	78
Beriev Be-6 and Be-10	89
Beriev Be-30	93
Ilyushin Il-12 and Il-14	96
Ilyushin Il-18 (piston engines)	111
Ilyushin Il-18 (airscrew-turbines)	113
Ilyushin Il-20/28	126
Ilyushin Il-62	128
Kamov Ka-15	136
Kamov Ka-18	139
Kamov Ka-25K	140
Kamov Ka-26	143
Mil Mi-1	150
Mil Mi-2 (V-2)	154
Mil Mi-4	160
Mil Mi-6	168
Mil Mi-8 (V-8)	178
Mil Mi-10 (V-10) and Mi-10K	183
Tupolev Tu-70	195
Tupolev Tu-104	198
Tupolev Tu-110	212
Tupolev Tu-114	215
Tupolev Tu-124	225
Tupolev Tu-134	233
Tupolev Tu-144	240

Tupolev Tu-154	246
Yakovlev Yak-12	252
Yakovlev Yak-16	257
Yakovlev Yak-18T	259
Yakovlev Yak-24	262
Yakovlev Yak-40	269
Appendix I—Aeroflot. Directorates, Main Bases, Areas of Responsibility, and Equipment—Summer 1967	277
Appendix II—Aeroflot. Equipment Used on International Services—Summer 1967 and Winter 1967–68	280
Appendix III—Aeroflot. Equipment Used on Soviet Federal Services—Summer 1966	282
Appendix IV—Routes Operated by Soviet-designed Aircraft in the Service of Non-Soviet Airlines—Summer 1967	290
Appendix V—World Records held by Soviet Transport Aircraft—1967	293
Appendix VI—Airline Fleets	296
Index	315

INTRODUCTION

Although the Soviet Union built its AK-1 single-engine transport monoplane as far back as 1923 and followed it with a variety of other transport types, it was not until after the 1939–45 war that the USSR became a major constructor of transport aeroplanes.

It has always been considered extremely difficult to acquire information concerning Soviet aircraft and to some extent this is true, but much has been published in the Soviet Union, particularly about the transport types, and in recent years very full information has become available soon after a new type has appeared. In some cases very detailed descriptions of design, construction and performance have been published for aircraft which were by no means ready for their trials, the Tu-154 being a case in point.

Many aspects of Soviet air transport are certainly regarded as State secrets and it is true that little information on airline operation is released abroad, but considerable detail can be discovered by research in Soviet Russian-language publications.

Air transport in the Soviet Union operates on a very large scale and is rapidly growing, and Aeroflot is steadily extending its international operations. Apart from Li-2s and some Czechoslovak taxi aircraft, the Aeroflot fleet consists of Soviet-designed aircraft. Many types have been built in very large numbers, and it must be agreed that in the overall world transport picture Soviet transport aircraft are of great importance.

It is for this reason that I decided to write this book covering all the Soviet transport aeroplanes and helicopters which have been produced since 1945. The earlier types have been described in my previous work *European Transport Aircraft since 1910*, and many of the later types also appear therein, but here I have been able to describe design features, construction and operational history at much greater length.

Most of the information has come from Soviet sources although I have been fortunate enough to study many of the aircraft at first hand and, in the case of the Il-12B, Tu-104A and Mi-10, I have flown in them.

In an introductory chapter I have set the scene by presenting the basic facts about the Soviet Union, in terms of geography, climate, population, industry and transport, and have also given a brief survey of Soviet air transport, its history and present scope. One still cannot simply ask Aeroflot for a timetable showing all its routes and schedules, and in fact it is possible that all its scheduled operations do not appear between the covers of any single publication. For this reason I have shown, in the appendices, the extent of Soviet Federal air services and, to underline the amount of work done by the major aircraft types, have listed these operations under the types of aircraft responsible for them. Also included are the scheduled operations by non-Soviet airlines operating Soviet equipment.

The Aircraft

As far as possible I have detailed for each type the design history, structural layout and features, accommodation and operational history. The technical data covers dimensions, weights and performance for as many variants as possible of each type. Unfortunately, the data presented is not consistent, and some aircraft are more fully covered than others.

For ease of reference aircraft have been arranged alphabetically by design bureaux, and within these sections the aircraft are shown in numerical sequence by type numbers. To some extent this method of presentation disturbs the strict chronological sequence but on balance the system chosen appears to be the most satisfactory.

All measurements are given in metric with English equivalents, but metric should be regarded as the most authoritative.

Speeds are shown in the data sections in km/h, knots and statute miles per hour and range figures are given in km and nautical miles, but the system has not been followed too rigidly in the text and in the case of FAI records speeds are shown in km/h and statute mph. Because of their similarity no attempt has been made to distinguish between cheval vapeur (cv) and horse power (hp). The term hp has been used throughout. 1,000 cv equals 986·32 hp.

Production

In most cases the exact number of Soviet transport aircraft produced is not known; nevertheless, as far as possible details of the civil fleets of Soviet-designed transports are listed in Appendix VI.

Cyrillic Alphabet

The Soviet Union uses the cyrillic alphabet, which has led to some confusion in identifying Soviet aircraft designations and registrations. The alphabet is shown here with the nearest Roman equivalents.

А	A	И	I	С	S	Щ	shch
Б	B	Й	Y	Т	T	Ъ	silent
В	V	К	K	У	U	Ы	ă
Г	G	Л	L	Ф	F	ЫЙ	ăy
Д	D	М	M	Х	kh	ь	'
Е	E or ye	Н	N	Ц	ts	Э	e
Ё	yo or o	О	O	Ч	ch	Ю	yoo
Ж	zh	П	P	Ш	sh	Я	ya
З	Z	Р	R				

JOHN STROUD

London, 1968

ACKNOWLEDGEMENTS

For help in the preparation of this work I record my appreciation of the kindness of the following individuals and organizations: *Aeroplane*, Air Canada, Air France, Jean Alexander, Aviaexport, *Aviation Magazine International*, *Aviation Week & Space Technology*, Bristol Siddeley Engines Ltd, Horst W. Burgsmüller, Charles Cain, R. A. Cole, ČSA—Československé Aerolinie, Cubana, *Flight International*, Ghana Airways, William Green, Heliswiss AG, M. J. Hooks, Indian Air Force, Japan Air Lines, *The Journal of Commerce and Shipping Telegraph*, W. B. Klepacki, Misrair, Norco Aviation & Industrial Equipment Ltd, Malcolm Passingham, Polskie Linie Lotnicze (LOT), Royal Nepal Airlines Corporation, SAS, E. Sekigawa and John W. R. Taylor.

I must give special thanks to my wife, Patricia, for all her help.

The response to requests for information for any project is normally disappointing. Of all the airlines which operate or have operated Soviet equipment only ČSA, Cubana, Ghana Airways, LOT, Misrair, Royal Nepal Airlines and Yemen Airlines responded to requests for facts.

The three-view drawings are by L. E. Bradford.

J.S.

SOVIET AIR TRANSPORT

In order to understand the development of Soviet transport aircraft it is necessary to know certain basic facts about the Soviet Union itself—its vast size, varied terrain, climate, distribution of resources and population, and surface transport. It is only in relation to this background that the reasons can be found for the design and production of some of the aircraft and for certain features of others.

The USSR

Including Estonia, Latvia and Lithuania, the USSR has an area of 22,402,000 sq km (8,649,172 sq miles) and a population of nearly 240 mn; by comparison, the United States of America, including Alaska and Hawaii, has an area of 9,363,389 sq km (3,615,207 sq miles) and a population of 200 mn.

The Soviet Union stretches from 20 deg E through 180 deg to 170 deg W and from beyond 80 deg N to nearly 35 deg S. Its European frontiers meet Norway, Finland, Poland, Czechoslovakia, Hungary and Rumania;

The Deruluft Fokker F.III RR 6 (360 hp Rolls-Royce Eagle VIII) being righted after a mishap at Kowno (Kaunas) on 19 June, 1924. Deruluft Fokker F.IIIs had opened a Königsberg–Kowno–Smolensk–Moscow service on 1 May, 1922. (*Courtesy Horst W. Burgsmüller.*)

11

Deruluft's ANT-9 URSS-D310 *Orel* (three 365 hp Wright Whirlwinds) at Riga, a call on the Königsberg–Leningrad route over which service began in the summer of 1934. (*Deruluft*.)

to the south are Turkey, Iran, Afghanistan, China and Korea; the eastern edge is washed by the Pacific, and the distance between the Soviet mainland and Alaska, across the Bering Strait, is only about 90 km (56 miles); the entire northern coast is bounded by the Arctic Ocean. Most of the northern ports are blocked by ice for nine months of the year, and off the northern coast are numerous islands, some never clear of the Arctic pack-ice, and these include Novaya Zemlya, Severnaya Zemlya, Zemlya Frantsa-Iosifa (Franz Josef Land), Novosibirskiya Ostrova (New Siberian Islands) and Ostrov Vrangelya (Wrangel Island).

Administrative Areas

The USSR has a very large number of administrative regions, although for the purposes of this book it is only necessary to consider the 15 Soviet Socialist Republics (SSRs) and the 20 Autonomous Soviet Socialist Republics (ASSRs), although the next levels include eight Autonomous Oblasts (AOs), 112 Krays or Oblasts and about 2,000 Rayons. Still further down the scale are about 40,000 rural Soviets.

The SSRs, with their capitals in parentheses, are: RSFSR–Russian Soviet Federated Socialist Republic (Moscow), Armenian SSR (Yerevan),

A Kalinin K5 (Pratt & Whitney Hornet), probably operated by Dobrolet and photographed somewhere in Central Asia.

Azerbaydzhan SSR (Baku), Byelorussian or White Russian SSR (Minsk), Estonian SSR (Tallinn), Georgian SSR (Tbilisi), Kazakh SSR (Alma Ata), Kirgiz SSR (Frunze), Latvian SSR (Riga), Lithuanian SSR (Vil'-nyus), Moldavian SSR (Kishinev), Tadzhik SSR (Dushanbe), Turkmen SSR (Ashkhabad), Ukrainian SSR (Kiev) and Uzbek SSR (Tashkent).

Of the 20 ASSRs, those that concern this work are Abkhaz and Adzhar in Georgian SSR, Kara-Kalpak in Uzbek SSR, Nakhichevan in Azerbaydzhan SSR, and the following all in RSFSR: Bashkir, Buryat Mongol, Chuvash, Dagestan, Karelo-Finnish, Komi, Mari, Mordovinian, Tatar, Udmurt and Yakut.

An Aeroflot Stal-3 (480 hp M-22) with flagman. Before the war Germany, too, used to follow railway practice in handling aircraft departures.

In addition to the administrative regions, there are a number of economic regions, some sharing Republic boundaries and others splitting the Republics. These economic regions are: Baltic, Byelorussia, Central Asia, Central Blackearth, Centre, Donets-Dnepr, Eastern Siberia, Far East, Kazakh, Moldavia, North-west, Northern Caucasus, South, South-west, Transcaucasia, Ural, Volga, Volga-Vyatka and Western Siberia. It will be found when studying the structure of Aeroflot that its administrative areas are, in general, based on the boundaries of SSRs, ASSRs and the economic regions.

By far the biggest of the Republics is the RSFSR which stretches right across the Soviet Union and embraces the whole of the northern and eastern coasts, second largest is Kazakh SSR and then comes Ukrainian SSR. Of the economic regions the biggest are Far East, Eastern Siberia, Kazakh, Ural and North-west. The ASSRs are for the most part administrative areas containing ethnic groups.

An aircraft which played an important role in providing air transport in remote areas was the U-2 biplane (100 hp M-11) which first appeared in 1927. The U-2, later redesignated Po-2, was produced in numerous versions and total production was reported as 20,000. The example shown is the U-2S2 ambulance.

Population

The RSFSR contains more than half the total population of the Soviet Union, more than 124 mn in 1964, and the Ukrainian SSR follows with more than one-sixth, at over 44 mn. Kazakh SSR had a population of $11\frac{1}{2}$ mn, Uzbek SSR just under 10 mn, Byelorussian SSR $8\frac{1}{4}$ mn, Georgian SSR and Azerbaydzhan SSR 4·4 mn each, and the other Republics had populations varying between 1·3 mn (Estonian SSR) and 3·2 mn (Moldavian SSR). In order of population density Moldavian SSR came first, 96·2 per sq km. Then came Ukrainian SSR with 74·3, Armenian SSR 69·4, Georgian SSR 63·3, Azerbaydzhan SSR 50·6, Lithuanian SSR 44·6 and Byelorussian SSR 40·7. The RSFSR had a population density of only 7·3 per sq km and Turkmen SSR was even less with 3·7. By far the greatest concentration of population is in the western part of the country between 20 and 50 deg E. East of this area there are populations scattered along the Trans-Siberian Railway, but the entire area to the north has a very low population density, with numbers of small remote settlements mostly along the rivers.

The largest Soviet cities, by populations officially estimated in 1962, are: Moscow (6,296,000), Leningrad (3,498,000), Kiev (1,208,000), Baku and suburbs (1,067,000), Gor'kiy (1,025,000), Tashkent (1,002,000), Khar'kov (990,000), Novosibirsk (985,000), Kuybyshev (881,000), Sverdlovsk (853,000), Donetsk (760,000), Chelyabinsk (751,000), Tbilisi (743,000), Dnepropetrovsk (722,000), Kazan' (711,000), Odessa (704,000), Perm (701,000), Rostov (661,000), Omsk (650,000), Volgograd (649,000), Saratov (631,000), Riga (620,000), Ufa (610,000), Minsk (599,000), Yerevan (583,000), Alma Ata (534,000) and Voronezh (516,000). At least 145 others have populations exceeding 100,000.

Terrain

To a large extent the Soviet Union can be considered in two sections, that which lies west of the Urals and that to the east of the range. The Urals run approximately north to south at 60 deg E. Except in the south, the area west of the Urals is generally flat and in the west and north has extensive marshlands. The area between the Urals and 90 deg E is also largely flat and swampy; but most of the terrain further east, right up to the north-eastern tip, is mountainous. In the southern regions there are large deserts and very high mountains. The country possesses three large areas of water, the Caspian Sea, the Aral Sea and, separating the main east coast from Kamchatka, the Sea of Okhotsk, and each of these has its effect on transport. The Ural mountains are not of great height; but in the southern areas of the Soviet Union the mountains are of Himalayan proportions.

Climate

A very large part of the country has a mean temperature of below freezing for half the year, and east of 85 deg E very few parts have less than 180 days a year with an average temperature above zero. At the other extreme there are areas in the south where temperatures average 20 deg C. (68 deg F.) or higher for up to 120 days a year, while in some areas very high day temperatures occur. In Verkhoyansk, in Yakut, January temperatures go down to a mean of minus 50 deg C. (-58 deg F.), and in Tashkent, at an altitude of 490 m (1,610 ft), the July mean temperature is 27 deg C. (80 deg F.).

Rainfall is heaviest in the west, parts of the south, the Urals and far east.

Before the war small flying-boats served areas inaccessible even to the U-2. The Sch-2, or more correctly Shch-2, amphibian (100 hp M-11) is seen here on the lake at the Tik-Guba settlement near Apatity in Soviet Lapland south of Murmansk.

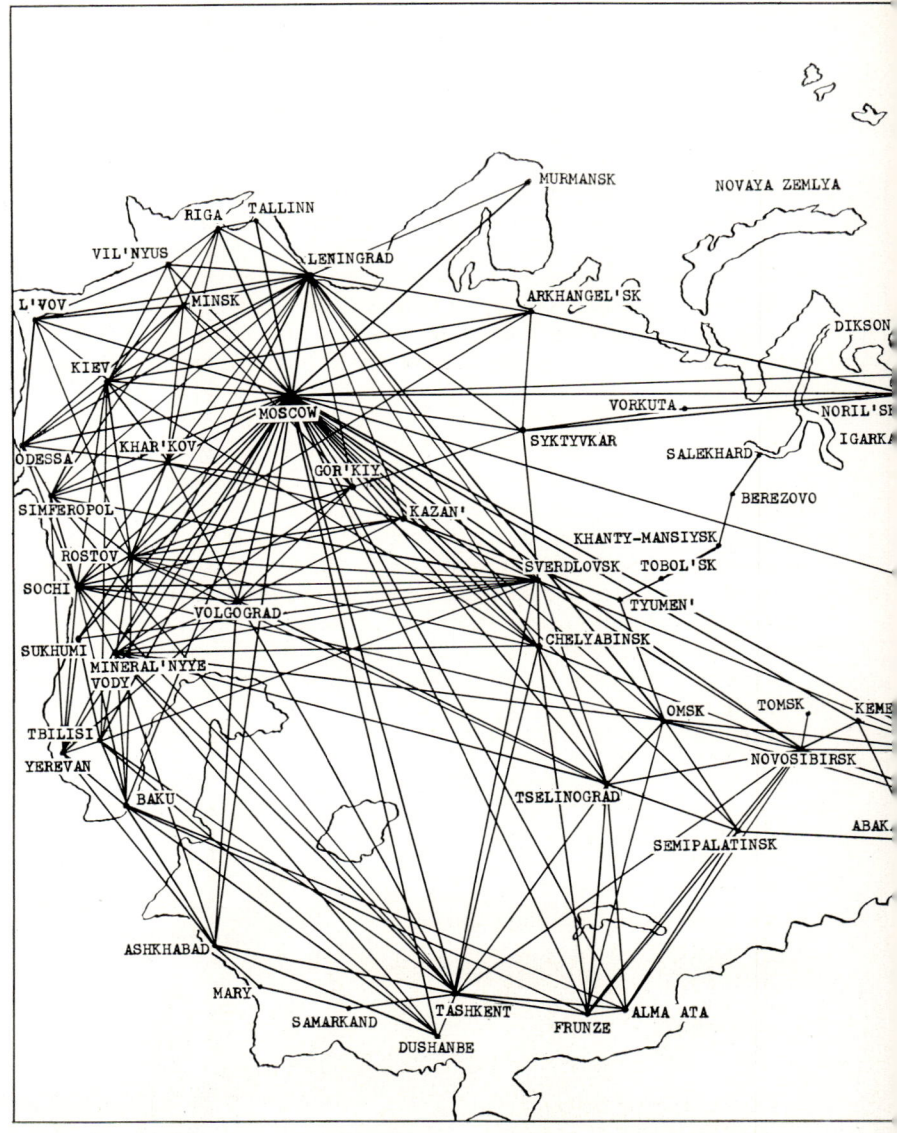

This map gives some impression of the extent of Aeroflot's domestic route system which in 1967 exceeded 500,000 km (310,690 miles) and served more than 3,500 cities, towns and settlements. The impossibility of showing the system in detail can be understood when it is realized that in 1966 there were more than 184,000 km (114,330 miles) of intra-Republic routes. The small scale of this map has made it necessary to omit many important cities, particularly in the western regions, and most of the local service routes,

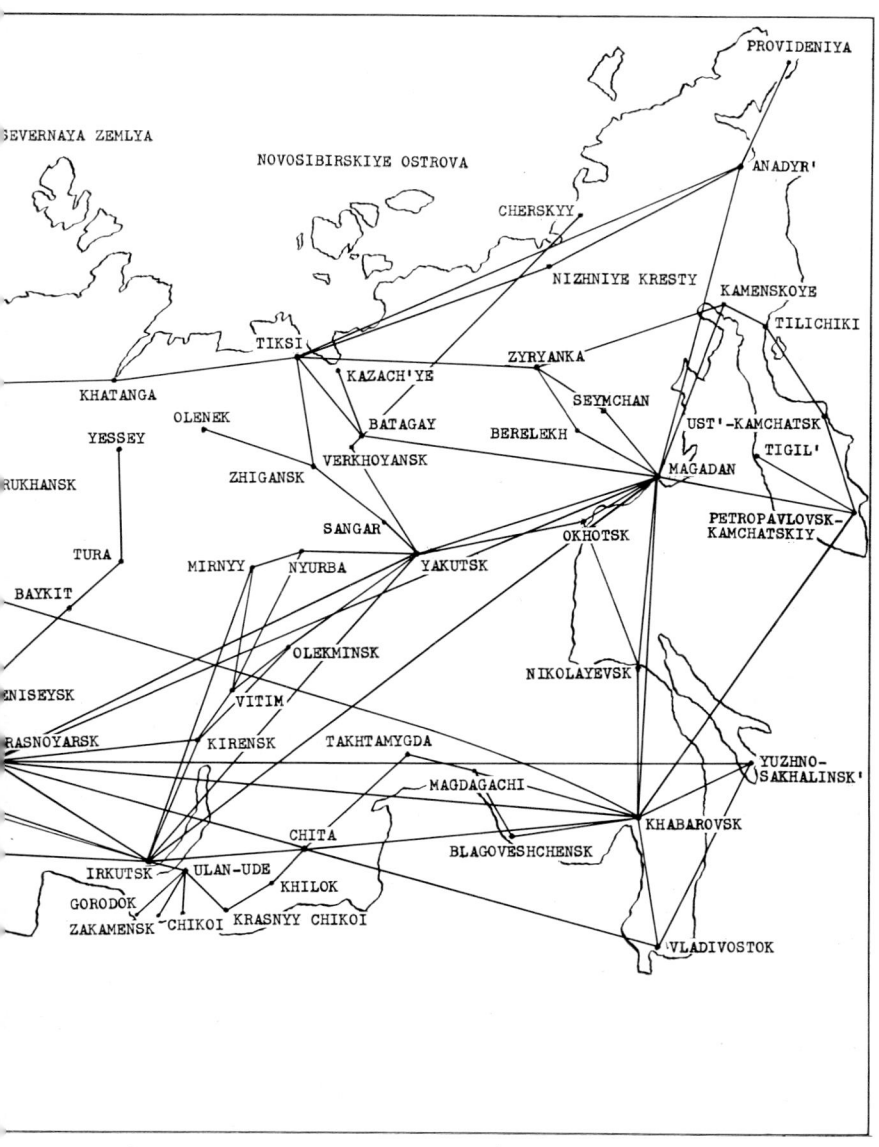

and it has only been possible to show less than 100 of the places served. Even in this simplified form, the density of operations to resorts such as Mineral'nyye Vody and Sochi is obvious. The concentration of routes in some areas combined with the vast extent of the Soviet Union makes it doubtful whether all of Aeroflot's routes could be shown on one map, unless it was truly enormous. A more detailed survey of Aeroflot operations appears in Appendix III. (© 1968 *John Stroud.*)

Certain limited areas have an average rainfall exceeding 100 cm (40 in) but very large areas have 10–20 cm (4–8 in) or even less.

In the far north snow lies for an average of 240 days a year, and even in the south there are areas where it lies for 20–40 days.

The western part of the Soviet Union suffers from general north European weather and the entire northern area is subject to fog because of proximity to the Polar ice.

Industry, Power and Agriculture

Forestry and timber represents an important part of the Soviet economy, approximately half of the total area of the country being covered by forests. Timber and wood pulp are major exports, and Arkhangel'sk, on the White Sea, is almost entirely devoted to this industry. Saw-milling and pulp and paper-making are largely concentrated in the western half of the country, but the handling of timber is widely distributed and pulp and paper-making is undertaken in the far east on Sakhalin.

Iron and steel is mainly concentrated in the Urals, with Chelyabinsk and Magnitogorsk as main centres, and in the Donbass where main industrial areas are Dnepropetrovsk, Krivoy Rog, Donetsk and Zaporozh'ye.

Like so many other things in the USSR, engineering is concentrated in the western part of the country, although Novosibirsk, Sverdlovsk and Tashkent, too, are all important engineering centres.

The chemical industry is concentrated in the west, the Urals, some southern areas and parts of Siberia, and the most important petroleum areas are Greater Baku on the Caspian, Second Baku where Kuybyshev is an important centre, and Emba on the north-east of the Caspian.

Considerable effort is being put into mineral exploration and much new industrialization is likely in the eastern regions; full exploitation

On 1 July, 1937, Aeroflot opened a Moscow–Riga–Stockholm service which was operated by Douglas DC-3s and ANT-35s. One of the latter, URSS-M131, is seen at Bromma Airport, Stockholm. (*AB Aerotransport.*)

It is believed that about 70 ANT-9 (PS-9) aircraft were used by Dobrolet and Aeroflot, most being the ANT-9-M-17 version with two 680 hp M-17 engines. One of the twin-engine aircraft is seen here in the Northern Caucasus in the autumn of 1942 when carrying supplies to the Soviet armies.

of minerals and industrial wealth is hindered by lack of transport facilities.

Main coalfields are in the Donbass, the Urals, Kuzbass and Irkutsk areas, with further deposits in the far east and Kirgiz.

The northern Ukraine and the blackearth region of central European Russia has some of the richest agricultural land to be found anywhere, but agriculture overall is severely limited by climate. European Russia is a major producing area, but outside this area climate restricts agriculture to a strip across southern Siberia, the Pacific coast and some southern areas such as the Asian Republics. Aviation plays an important part in the increasing and protection of crops and is also used for their transport.

Transport

The Soviet railway system is believed to measure about 130,000 km (80,780 miles), compared with about 350,350 km (217,700 miles) in the United States. In the Soviet Union there are about 0·52 km (0·32 miles) per 100 sq km (38·6 sq miles) of territory and 54 km (33·5 miles) per 100,000 people; whereas in the United States the figures are 4·7 km (2·9 miles) per 100 sq km and 222 km (138 miles) of line per 100,000 people. Almost the entire Soviet railway system is west of the Urals apart from the Trans-Siberian and Turksib Railways. Some 45 per cent of all goods are moved by train, and railway construction is still being undertaken. It was planned, for example, that about 60,000 km (37,280 miles) of new track should be built in the period 1960–80.

Although the road system is extensive, the mileage of asphalt or concrete surfaced roads does not greatly exceed 100,000 km (62,000 miles). Very many roads are impassable or at least difficult during winter and spring.

Rivers and canals play quite an important part in the Soviet transport system but, although the total length of navigable waterways equals that of the railway system, water transport handles only about one-tenth of the total cargo traffic. Ice is a great problem, few main rivers remaining open throughout the year, and there is also a draught problem because of fluctuations in rainfall. Canals are still being constructed, and passenger and mail services are operated on the waterways in many parts of the country.

Aeroflot

Most Soviet civil air operations are worked under the title Aeroflot. This name is a popular contraction from Grazhdanskaya Vozduzhnaya Flot, the Civil Air Fleet, an organization recently superseded in overall responsibility by the Ministry of Civil Aviation. Aeroflot has recently been celebrating 50 years of existence, but this is a rather elastic interpretation of its history, for Aeroflot only came into being in 1932 when all Soviet air services came under the Chief Administration of the Civil Air Fleet.

The Chief Air Fleet Administration was founded on 24 May, 1918; in 1921 or 1922 there appear to have been some experimental passenger and mail flights over a route from Moscow to Sevastopol' via Tula, Orel, Kursk and Khar'kov by a converted Sikorsky *Il'ya Muromets* type bomber, but no regular service was operated; and in 1923 three Soviet air transport undertakings were established.

In 1922 air services with Soviet participation had been started when on 1 May the joint Soviet–German airline Deruluft opened the Königsberg–Kowno–Smolensk–Moscow route with Fokker F.IIIs. This operation was later expanded to have Berlin as its western terminal, and a Königs-

During the war large numbers of Po-2 biplanes were used for ambulance work. Three camouflaged examples appear in this photograph. In the original print it can be seen that the nearest aircraft carries a civil registration, the name Aeroflot forward of the cockpit and a red cross aft of the light-coloured fuselage band.

berg–Tilsit–Riga–Tallinn–Leningrad service was also operated. Later, Dornier Merkurs and Rohrbach Rolands were used, but for several years the ANT-9 and Junkers-Ju 52/3m were mainly responsible for Deruluft operations, although in 1937 some Deutsche Lufthansa Heinkel He 111s were employed on the Berlin–Danzig–Königsberg sector of the Berlin–Moscow route.

Backbone of Soviet postwar air transport was the Li-2, Soviet version of the DC-3. Aeroflot's Li-2 SSSR-L1015, seen here at Malmi Airport, Helsinki, on 6 June, 1949, had dark green upper surfaces with high-gloss finish. (*John Stroud.*)

Dobrolet was founded in March 1923, and the Red Air Fleet provided the airline with aircraft which included a de Havilland 34 and a Vickers Vimy Commercial. A service was opened between Moscow and Nizhniy Novgorod (now Gor'kiy), and in the Soviet Union this is now regarded as the start of the country's air transport system. In July 1924 the route was extended to Kazan' and AK-ls and Junkers-F 13s were used. Services were established from Moscow to Batumi, Khar'kov, Kiev, Odessa, Orel and Tbilisi, while in Central Asia, Khiva–Bokhara–Dushanbe and Tashkent–Alma Ata services were opened. Also founded in 1923 was the Ukrainian airline Ukrvozdukhput which operated Dornier Komets and opened services between Khar'kov and Kiev and Khar'kov and Odessa. It is thought that later the airline employed the Kalinin K4.

A third pioneer concern was Zakavia which was also founded in 1923 and operated services from Tbilisi to Baku and Yerevan. Zakavia was taken over by Ukrvozdukhput in 1925, and in 1930 the Ukrainian airline and Dobrolet were reorganized as Dobroflot. Further reorganization in 1932 brought Aeroflot into being. There is some evidence to suggest that for an intermediate period in 1933–34 the name Transaviatsia was used.

In 1933 Glavsevmorput (Administration of Northern Sea Routes) was formed and operated flying-boats along certain rivers and then, later in the 1930s, began some east–west operations.

Flying in the far north was controlled by Aviaarktika, a development organization, and its duties included making deliveries to research groups and weather stations on the ice cap and supplying otherwise inaccessible

Until replaced by An-2s and Yak-12s, Po-2s continued to serve remote areas. This ski-equipped example is seen at Ust'-Kara in Nenets, on the Arctic coast, in 1947.

parts of northern USSR. Glavsevmorput operations may have become the responsibility of Aeroflot, and Aviaarktika appears to have become Polyarnaya Aviatsiya (Polar Aviation) or the Polar Directorate of Aeroflot.

In 1932, when Aeroflot was formed, some 27,000 passengers and less than 900 tons of freight and mail were carried. By 1933 there was a main trunk air route with regular services stretching from Moscow to Irkutsk via Kazan', Sverdlovsk, Novosibirsk and Krasnoyarsk. North from Irkutsk there were routes to Olekminsk and Yakutsk, and to Bodaybo. In the far east there were regular services between Vladivostok and Okha via Khabarovsk and an irregular link between Irkutsk and Khabarovsk.

In the western part of the country Moscow was linked to Stalinogorsk (now Novomoskovsk), Stalingrad (now Volgograd), Elista, Khar'kov, Kiev, Odessa, Zhdanov, Rostov, Adler/Sochi, Mineral'nyye Vody, Baku, Kuybyshev and Aktyubinsk. There was a service between Baku and Tbilisi and irregular services between Aktyubinsk and Tashkent from where a small network served Samarkand, Urgench, Tashauz and a few other points. An isolated route linked Karaganda, Balkhash and Alma Ata, and another was between Tyumen and Obdorok in the far north. There were two branches off the main east–west trunk route, from Sverdlovsk to Kustanay and Novosibirsk to Norchansk. There was a mail operation between Moscow and Leningrad and two international routes, from Tashkent to Kabul and Ulan-Ude to Ulan Bator.

By 1935 considerable expansion had taken place, 111,000 passengers were carried in the year and more than 11,000 tons of mail and cargo. Also by that time Aeroflot was administratively divided into 12 regional managements: Azov–Black Sea, Caucasus, Central Asia, East Siberia, Far East, Moscow, North, Trans-Caucasus, Ukraine, Ural, Volga and West Siberia.

In 1940, the last full year before the German attack, Aeroflot had a route network of some 146,300 km (90,906 miles) and carried nearly 359,000 passengers and about 45,000 tons of mail and freight.

During the war certain essential routes to Siberia were kept open, as were some of those to the Arctic, but Aeroflot's main activities were in support of the armed forces. Twenty-one of Aeroflot's pilots were made Heroes of the Soviet Union and about 15,000 Aeroflot crew members were awarded decorations for war service. Some 2,300,000 passengers were carried of whom hundreds of thousands were troops, 4,500,000 hr were flown and more than 400,000 tons of war cargo carried. Over 40,000 flights were made beyond enemy lines and about 900 tons of leaflets dropped. A considerable amount of ambulance flying was undertaken, mostly by single-engine U-2 (Po-2) biplanes.

Aeroflot after the War

When peace returned Aeroflot's principal task was the maintenance and re-establishment of essential services, and for some years the operations were of necessity conducted under austerity conditions in terms of aircraft, airports and equipment in general.

Li-2s, still in camouflage green paint but sometimes with high-gloss finish, formed the backbone of the fleet, with the Po-2s continuing to serve the remote areas. In the autumn of 1947 the first Soviet postwar transport, the Il-12, entered service, which was to play an important part in Aeroflot's development even though it took 33 hr to fly from Moscow to Vladivostok, making nine landings on the way.

In 1946 the Supreme Soviet decided that the country's air transport network should be extended to 175,000 km (108,740 miles) and that Aeroflot should be provided with new aircraft. This expansion and re-equipment has been a continuous process ever since.

By 1950 the route network measured 300,500 km (186,720 miles) and during that year Aeroflot carried 1,600,000 passengers and 181,500 tons of cargo and mail. In addition, 3,400,000 hectares (about 8·5 mn acres) of agricultural land received attention from aircraft. In 1955 the network

A camouflaged Aeroflot Li-2 at Mineral'nyye Vody soon after the war. The flagman is still in evidence.

In 1947 Il-12s began operation. An Il-12B, with dorsal fin and early postwar Aeroflot colour scheme, is seen at Malmi Airport, Helsinki.

measured 321,500 km (199,769 miles), 2,500,000 passengers were carried and 258,700 tons of mail and cargo, while agricultural aircraft treated 9,900,000 hectares (about 24·75 mn acres).

During the period 1950–55 the Il-14 entered service, and Aeroflot began using helicopters for various duties. The most important decisions taken during this period were those leading to the building of the Tu-104 turbojet and An-10 and Il-18 airscrew-turbine aircraft, for it was these which began to bring about mass air travel and which were to make Aeroflot the biggest single passenger air carrier in the world.

In September 1956 the Tu-104 entered service, between Moscow and Irkutsk, cutting the journey time from nearly 18 hr to $6\frac{1}{2}$ hr. This aircraft, although not particularly economic since it could accommodate only 50 passengers, had spectacular performance and was a major achievement because with it the USSR jumped straight from the small twin-engine Il-12 and Il-14 to the turbojet without any intermediate civil airliner; elsewhere the turbojet transport followed a succession of sophisticated twin and four-engine types. It should be remembered, too, that the Soviet Union had little or no experience in building large pressure cabins, and it was probably for this reason that the Tu-104 has a pressure bulkhead between the passenger area and the flight deck, as well as having individual oxygen supply for each passenger.

The first Soviet jet transport made its first flight in June 1955. This was the Tu-104 which entered service in September 1956. The prototype is seen here at London Airport (Heathrow) in March 1956.

In 1958 the 100-passenger Tu-104B entered service on the Moscow–Leningrad route and in 1959 the An-10s and Il-18s began regular operation. The new equipment caused a spectacular rise in traffic, and during 1959 Aeroflot carried 12,300,000 passengers over a route network which had grown to 355,400 km (220,870 miles).

April 1961 saw the very big airscrew-turbine Tu-114 enter service on the Moscow–Khabarovsk route and, later, this type opened Aeroflot's first transatlantic services, to Cuba and Montreal. October 1962 was an important date for it was during that month that turbine-powered aircraft began operating over short routes when the turbofan Tu-124 and airscrew-turbine An-24 entered regular service. By the end of that year Moscow and the capitals of all the Republics were linked by turbine-powered aircraft.

Vnukovo Airport, Moscow, in about 1957 with the Tu-104As SSSR-L5428 and SSSR-L5438, and centre, Tu-104 SSSR-L5419. In the background are numbers of Il-14s and cargo versions of both the Il-12 and Il-14.

Throughout this period the traffic growth rate was astronomical. In 1965 Aeroflot carried 42 mn passengers and 1 mn tons of cargo and mail over a route network grown to 500,000 km (310,690 miles). Aerial work over farms and forests covered an area of 55 mn hectares (about 137·5 mn acres).

Sometime earlier it had been decided that the air should be the main means of passenger transport in the USSR, and there was no question that this was coming true, for 1966 saw Aeroflot's passenger traffic increase to 47,200,000 while its 1967 target figure of 53 mn was surpassed. It is estimated that by 1970 Aeroflot will carry 75 mn passengers in the year with 35 mn of them flying on local services.

To cater for this enormous traffic it has been necessary to produce new generations of aircraft. During 1967 the twin-jet Tu-134 and four-jet Il-62 entered service, as well as the twin-turbine Mi-2 and Mi-8 helicopters. To replace the An-10s, Il-18s and Tu-104s, the much bigger trijet Tu-154 is under development. For local services the airscrew-turbine Be-30 and turbofan Yak-40 are undergoing flight testing, and the 80–100 seat Mi-6P helicopter is also on test. To come is the Soviet supersonic airliner–the Tu-144.

The following table of comparative air and rail journey times and fares goes far to explain the boom in Soviet passenger air travel.

Routes from Moscow	Distance		Time		Fare (roubles)	
	km	miles	air	rail	air	rail
Aktyubinsk	1,475	(916)	6 hr 30 min	(30 hr 40 min)	27	(26·50)
Alma Ata	3,095	(1,923)	5 hr 10 min	(67 hr 35 min)	54	(44·60)
Askhabad	2,775	(1,724)	5 hr 45 min	(81 hr 04 min)	46	(50·10)
Baku	1,985	(1,233)	3 hr 25 min	(41 hr 50 min)	34	(32·20)
Blagoveshchensk	6,070	(3,771)	11 hr 10 min	(147 hr 56 min)	102	(81·10)
Bratsk	4,095	(2,544)	8 hr 05 min	(92 hr 46 min)	68	(48·40)
Chita	4,840	(3,007)	9 hr 10 min	(102 hr 44 min)	81	(64·70)
Dushanbe	3,020	(1,876)	5 hr 15 min	(83 hr 18 min)	54	(50·10)
Frunze	2,895	(1,798)	5 hr 05 min	(73 hr 25 min)	52	(42·60)
Gor'kiy	430	(267)	50 min	(6 hr 18 min)	11	(11·50)
Irkutsk	4,220	(2,622)	6 hr 45 min	(81 hr 37 min)	72	(55·70)
Khabarovsk	6,900	(4,287)	8 hr 20 min	(149 hr 15 min)	108	(89·50)
Kiev	740	(459)	1 hr 10 min	(11 hr 55 min)	15	(17·40)
Krasnoyarsk	3,555	(2,208)	5 hr 30 min	(62 hr 16 min)	58	(44·60)
Kuybyshev	960	(596)	1 hr 15 min	(16 hr 00 min)	18	(18·60)
Leningrad	655	(407)	1 hr 05 min	(7 hr 45 min)	13	(14·20)
Mineral'nyye Vody	1,360	(845)	1 hr 55 min	(26 hr 16 min)	26	(27·30)
Novosibirsk	2,905	(1,805)	4 hr 35 min	(47 hr 46 min)	49	(37·40)
Omsk	2,305	(1,432)	3 hr 10 min	(39 hr 50 min)	40	(32·20)
Samarkand	3,015	(1,873)	7 hr 10 min	(67 hr 56 min)	52	(42·60)
Semipalatinsk	2,850	(1,770)	6 hr 15 min	(78 hr 55 min)	50	(42·60)
Sochi (Adler)	1,385	(860)	2 hr 05 min	(29 hr 53 min)	26	(28·00)
Tashkent	2,850	(1,770)	4 hr 35 min	(56 hr 40 min)	48	(37·40)
Yerevan	1,920	(1,193)	3 hr 15 min	(51 hr 10 min)	34	(33·10)

It will be seen that in some cases air fares are cheaper than rail.

Although much of Aeroflot's activity caters for the business traveller, there is now a fantastic amount of holiday travel, particularly to Black Sea resorts. There are high-frequency services between Moscow and Adler, for Sochi, and in recent years more and more cities have been given direct air services to satisfy this urge to reach the sun and the sea. The traffic converging on these Black Sea resorts and on places such as Mineral'nyye Vody and Simferopol is equivalent to the annual mass exodus of Scandinavians who, after their long dark winter, seek the sun in Spain and Italy.

The Tu-104s were joined in Aeroflot service in 1959 by the airscrew-turbine Il-18s. Three Il-18s are seen lined up for departure from Baku, Azerbaydzhan. The signs in the foreground bear flight numbers and, from left to right, show the destinations: Yerevan, Ashkhabad, Tashkent, Kuybyshev, Kiev, Simferopol, Khar'kov, Moscow, Leningrad, Volgograd, Krasnovodsk and Adler.

Cargo

Although the statistics leave no doubt that Aeroflot is a system for the mass transport of passengers, it plays an important role in carrying a very wide range of cargo. Cargo has always been given high priority on scheduled passenger services but the volume of scheduled all-cargo services is being steadily increased. Li-2s, Il-12s and Il-14s have all been responsible for cargo operations and more recently a network of scheduled services has been built up with An-12s, and this type is even used for the carriage of fruit and animals. Undergoing trials is the very big An-22 which can

Although the double doors of lend-lease Douglas C-47s were removed and replaced by single doors on the starboard side, some of Aeroflot's Il-12s, quite soon after their introduction, were converted to freighters and fitted with double doors on the port side. Cargo for Kuybyshev is seen being loaded into the Il-12 SSSR-L1458.

An Aeroflot survey Li-2.

carry heavy equipment, operate from unpaved runways and handle loads of up to nearly 80 tons. Helicopters, too, are used for cargo carriage and as cranes for heavy lifting.

On many routes out of Moscow priority is given to the more than 200 sets of newspaper matrices which go every day to some 38 cities where over 27 mn newspapers are printed and delivered for publication on the same day as Moscow.

In addition to the extensive domestic cargo operations, Aeroflot operates a regular An-12 cargo service to Djakarta and, in association with Air France, a weekly Moscow–Riga–Paris cargo service.

Agricultural and Ambulance Work

Aeroflot is also responsible for a vast amount of agricultural work, forest protection and ambulance flying. Some details of this will be found in the chapters describing the aircraft used, notably the An-2, the Yak-12 and the smaller helicopters. Geological survey and photography are also tasks on which Aeroflot is employed. The Soviet Union has shown some interest in the potential of the balloon for the lifting of timber from mountain sites and it is believed that some experiments have been made.

This photograph of an Mi-1 spraying shows the pattern of distribution caused by the rotor vortex.

Two Polar Aviation Il-12Bs at Mirnyy in Antarctica.

Polar and Cold Weather Operation

Comparatively little is known about Soviet Polar aviation but it can be stated with certainty that Aeroflot, and in particular its Polar Directorate, has unrivalled experience of operation under Arctic conditions.

Regular services are operated throughout the northern parts of the Soviet Union and some 3,000,000 km (1,864,000 miles) are flown annually on ice reconnaissance and the escorting of ships through the ice. Supplies are flown in to the northern settlements and fish is carried from the Arctic shores. Arctic and Antarctic scientific parties are supplied by air and the work involved has frequently necessitated operating from drifting ice-floes.

Northern operations have forced the use of skis on the smaller aircraft, and trials have been made with the An-12 on heated skis. All the main

The An-12 SSSR-04366, on skis, arriving at the Soviet Vostok base in Antarctica. This An-12 had an orange-painted fin and rudder with a black and white penguin which can be seen just above No. 4 engine.

types of Soviet transport aircraft have full deicing and anti-icing systems and this also applies to the rotors of helicopters.

Pilots flying in the northern areas receive the highest rates of pay and there is extra pay for the first landing on ice on each flight.

The Size of Aeroflot

Regarded simply as an airline, Aeroflot is enormous and is certainly carrying more passengers and cargo than any other air carrier anywhere. Its operations can best be compared with those of the US airlines which serve a population of approximately the same size although concentrated into a much smaller territory. In 1967 the US carriers handled well over 100 mn passengers or approximately double the number flown by Aeroflot.

Aeroflot is vast and large numbers of its aircraft are to be seen at all main Soviet airports. This 1961 view of just a corner of Vnukovo Airport shows seven Aeroflot Il-18s, five Tu-104s and a Tu-124. The nearest aircraft, SSSR-42475 and SSSR-42401, are Tu-104Bs. Also visible are Il-18s of Ghana Airways and Malév.

On this basis it would be reasonable for Aeroflot to have less aircraft than the combined US airlines, especially as Soviet load factors are believed to be higher, but all indications point to Aeroflot achieving much lower fleet utilization. Also Aeroflot carries very large numbers of passengers in small aircraft operating local services. The comparison is also complicated by the much greater extent of the US airlines' international operations, by Pan American World Airways, Northwest Airlines and TWA in particular.

It is impossible to get information on the number of Aeroflot aircraft, but it is not unreasonable to suggest the following approximate round figures: 150 An-24s, 200–300 An-10s and An-12s, 400 Il-18s, 100 Tu-124s, 30 Tu-114s and 200 Tu-104s. Il-62s and Tu-134s are in service in small but increasing numbers. To these figures must be added very large numbers of An-2s and small helicopters. There are some An-14s, a large number of Mi-4s, small numbers of Mi-6s and Mi-10s, and probably quite a lot of Yak-12s. It is possible that main fleet units come to somewhere around 1,000, and that, including small aircraft, the fleet numbers not less than

3,000. Very many of the smaller aircraft are engaged on duties which are not undertaken by US airlines although An-2s and Yak-12s are to some extent fulfilling the functions of the US third-level operators.

Approximate figures for US turbojet-powered fleets at mid-1967 were 947, comprising 232 Boeing 707s, 130 Boeing 720s, 268 Boeing 727s, 46 Convair CV-880s, 19 Convair CV-990s, 125 Douglas DC-8s, 52 Douglas DC-9s, 56 BAC One-Elevens and 20 Sud-Aviation Caravelles. The main fleets of airscrew-turbine and piston-engined aircraft totalled 780, comprising 176 Convair-Liners, 92 Douglas DC-3s, 132 Douglas DC-6s and DC-7s, 92 Fairchild F-27s and FH-227s, 41 Lockheed Constellations and Super Constellations, 98 Lockheed Electras, six Lockheed Hercules, 75 Martin

In 1967 Il-62s began replacing Tu-114s. The prototype or pre-production Il-62 SSSR-06176 is seen with the Tu-114 SSSR-76466.

4-0-4s, 12 Nord 262s, 55 Vickers-Armstrongs Viscounts and three Japanese YS-11s. There were also numerous other small aircraft, mostly in Alaska, and the fleets of the third-level operators and helicopter airlines.

Aeroflot pilots have been reported to number about 20,000, but no accurate estimate can be made of the total staff.

Aeroflot Administration

It is known that by 1935 Aeroflot had 12 regional managements but details of its various administrative reorganizations are not known. However, it is known that in 1964 there were 18 Territorial Directorates (Territorialnoe Upravlenie–TU), nine Aviation Groups (Otdyelnaya Aviagruppa–OA) and two special organizations.

The Territorial Directorates and their headquarters were: Azerbaydzhan (Baku), Byelorussian or White Russian (Minsk), Far East (Khabarovsk), Georgian (Tbilisi), Kazakh (Alma Ata), Krasnoyarsk (Krasnoyarsk), Moscow (Moscow-Bykovo), Moscow Transport Directorate (Moscow-Vnukovo), Volga (Kuybyshev), Northern Caucasia (Rostov), North (Leningrad), Ukrainian (Kiev), Ural (Sverdlovsk), Uzbek (Tashkent),

Eastern Siberia (Irkutsk), Yakut (Yakutsk), Western Siberia (Novosibirsk) and Turkmen (Ashkhabad).

The Aviation Groups and their headquarters were: Armenian (Yerevan), Estonian (Tallinn), Kirgiz (Frunze), Latvian (Riga), Lithuanian (Vil'nyus), Magadan (Magadan), Moldavian (Kishinev), Syktyvkar (Syktyvkar) and Tadzhik (Dushanbe).

The special organizations, both Moscow based, were TUMVL responsible for international operations, and Polar Aviation.

There was a further breakdown into smaller administrative areas, for example, the Moscow Transport Directorate had 15 separate divisions.

Each Directorate or Group was responsible for the operation of services and other air activity in its own regional area and for certain Federal operations. For example, the Ukrainian Directorate, apart from being responsible for all operations within the Ukraine, undertook the working of all services from Kiev, Khar'kov, L'vov, Odessa and Simferopol to Moscow as well as some to Leningrad, Mineral'nyye Vody and Sverdlovsk.

Some Federal services were operated in pool by several Directorates, as was the trans-Siberian route between Moscow and Khabarovsk.

The Directorates and Groups appear to have been autonomous with their own fleets, headquarters, maintenance organizations and staff, but if required, aircraft were exchanged between Directorates.

Subsequent to 1964 certain changes were made. The Aviation Groups, TUMVL and Polar Aviation have all become Directorates, the term Territorial is no longer used, and the Syktyvkar Aviation Group is now the Komi Directorate. Responsibilities would appear to be the same. There is also a unit known as 235 Division. This operates An-24s, Il-18s and Tu-124s over a small number of routes but its purpose is not known.

Each Directorate is believed to carry out some training but it is thought that most is undertaken by organizations serving the whole of Aeroflot.

Safety

There is a complete absence of safety statistics for Soviet air transport, nevertheless, there is no reason to believe that operations in the USSR are any less safe than elsewhere. Crew training is thorough and Aeroflot has an advantage over most western airlines because of its widely varied fleet and range of work. Crews gain experience on small aircraft and progressively work their way up to command of the main types, a system not possible for airlines having fleets consisting solely of one or two types of large turbojet transports.

Transport aircraft produced in recent years all incorporate safety equipment and are adequately protected against fire, and pressurized aircraft, since the Tu-104 and Tu-114, have been tank tested to ensure fuselage integrity.

Although the Soviet Union is not a member of ICAO and neither is Aeroflot a member of IATA, most of the aircraft design features and performance comply with ICAO standards, and in general, Aeroflot, on its main operations, follows IATA practice.

All new types of aircraft undergo long periods of testing and then operate cargo services before being cleared for passenger carriage.

To a large extent Aeroflot is operating to airports and aerodromes which do not yet have full radio, radar and lighting aids, but progress is being made, autoland and semi-autoland is being developed, and meanwhile Aeroflot operates to higher weather limits than in the west.

Great attention is given to crew health, duty hours and rest periods.

The 4,250 hp Ivchenko AI-20M airscrew-turbines of the Il-18D SSSR-75581.
(*John Stroud.*)

Aircraft Design, Production and Marketing

Design of aircraft is undertaken by design bureaux which bear the names of the designers which head them, such as Antonov, Ilyushin, Mil, Tupolev and Yakovlev. Specifications are normally issued to only one bureau, but in some cases two design teams may work to provide their own solutions to a problem, as in the case of the Il-18 piston-engine transport and the Tu-70.

Normally a small number of prototypes are constructed for flight, dynamic and fatigue testing. Factory trials and State trials are followed by series production, if the aircraft is approved, and in some instances there are pre-production batches.

Production is allocated to the factories with available capacity. For example, it is believed that the prototypes and first production batches of the An-24 were built at Kiev, home of the Antonov design bureau, but, more recently, their production has begun at another factory at Ulan-Ude in the east near Lake Baykal.

Export of Soviet aircraft is the responsibility of Aviaexport, which is based in Moscow and is also the publisher of all Soviet aircraft sales brochures and publicity material. Aviaexport arranges participation in the Paris Aero Show and other exhibitions and organizes demonstrations such as that of the Mi-10 at Gatwick in March 1967.

Reported sales up to the end of 1967 include 650 Mi-4s, 30 Mi-6s and 90 Il-18s.

Engine Time between Overhauls

The Soviet Union has not found it easy to sell transport aircraft to airlines which are free to buy in world markets. Part of this reluctance was initially due to a Soviet unwillingness to provide full information and to the Russians' lack of commercial experience, but a most important factor was the very poor showing of Soviet aero-engines.

In 1961, after two years in regular service, the AI-20 airscrew-turbine in the Il-18 had an approved TBO of only 500 hr, and some of the early operators of Il-18s got only 50–100 hr out of the engines. However, the Russians were aware of the problem, it was even the subject of a speech by Krushchev, and by 1962 the overhaul period was increased to 750 hr. By 1963 AI-20s were cleared for 2,000 hr and in 1965 the AI-20K was achieving 4,000 hr, with some engines running to 6,000 hr by 1967.

When the Il-62 entered service in 1967 its NK-8-4 turbofans were approved for 3,000 hr.

Aircraft Features

Most of the special features of Soviet aircraft will be found described in the appropriate chapters, but some of these features require mention. Poor aerodromes have been responsible for Soviet insistence on good

The Mi-4 SSSR-31479 producing its own local snowstorm as it lands at Khabarovsk.

take-off and landing performance, and the use of low-pressure tyres and multi-wheel undercarriages is dictated by the need to operate from unpaved surfaces. It is also for this reason that tyre pressures of some aircraft can be changed in flight. By introducing variable tyre pressure, utilization has been increased with consequent economic gain.

Concentration on deicing and anti-icing systems, particularly for helicopters, has been dictated by weather conditions, and it is for the same reason that much development work has gone into the provision of autopilots and blind flying equipment for helicopters.

There is some indication that the Soviet Union is ahead in the fitting of angle of attack indicators to jet transports. British pilots have expressed the view also that Soviet aircraft radar, at least in aircraft with chin-positioned scanners, has better definition.

Aircraft instruments have in the past been inferior to those in the west but these deficiencies now appear to have been overcome, although some of the cabin furnishing materials could still be improved.

Strangely, Soviet jet transport aircraft have not, until the Il-62, been equipped with thrust reversers, and several types employ braking parachutes. Soviet turbojet aircraft have also lacked noise suppressors although these will be fitted to the NK-144 engines in the supersonic Tu-144.

Helicopters

Russia has a long history of rotary-wing development, but it was not until some years after the 1939–45 war that the first production helicopters began to appear. The first type produced in quantity was the Mil Mi-1, and since that time the Mil design bureau has become the principal source of Soviet helicopters. The Mi-1 was followed by the Mi-4 which was produced in large numbers and had the distinction of being the first Soviet helicopter to go into scheduled service. Then came the twin-turbine Mi-6 and Mi-10 which are still by far the biggest helicopters in the world. Turbine-powered successors to the Mi-1 and Mi-4 were produced as the Mi-2 and Mi-8, and the Mil team claims to have produced more helicopters than any other single organization—a claim which may well be true.

In the Soviet Union the helicopter is an extremely important vehicle, providing transport of passengers, cargo and heavy equipment to otherwise inaccessible areas, and also undertaking ambulance work and a wide range of agricultural duties.

Kamov helicopters are in service with the Soviet Air Force and with Aeroflot, but in civil operation they are mainly used for agricultural work.

There are almost certainly more helicopters at work in the USSR than in any other country, and the Soviet Union quite definitely leads in design, construction and operation of the very big helicopter. Soviet designers appear to have been the pioneers in providing rotor blades with ice protection.

Development of the rotary-wing aircraft as a short-haul city-to-city

The very big helicopter is playing an important role in the Soviet economy and has also undertaken work outside the USSR. An Mi-6 is seen here working for Heliswiss when it carried a cable-railway cabin to a mountain station in the Bernese Oberland in 1966. (*Courtesy Heliswiss.*)

vehicle is almost certain, and the Soviet Union offers much scope for the operation of this type of helicopter or compound aircraft.

The Mi-2, Mi-8 and Mi-10 are also referred to by the designations V-2, V-8 and V-10, the V standing for vertolet, Russian for helicopter.

Records

The USSR is very conscious of the capabilities of its aircraft for setting records for speed, distance, altitude and load carrying. Almost every transport aeroplane and helicopter produced since the war has established a series of records, frequently broken again by subsequent types, and those that still stood in 1967 are detailed in Appendix V.

Airports

The number of Soviet airports and civil aerodromes is not known, but within the country some 3,500 cities, towns and settlements are served by regular air services. In addition, ambulance aircraft operate to hundreds of small fields and agricultural aircraft work from strips throughout the country.

The aerodromes are situated in a wide variety of terrain and in some places at considerable elevation, many being simply dirt or grass surfaced and some snow-covered for many months of the year.

An enormous airport and aerodrome improvement programme has been undertaken in recent years but it will be a considerable time before all can be brought to a universally high standard.

In the period 1966–70 plans call for the building of 35–40 airports of what are called all-Union significance, while no less than 200 are to be constructed for local services. Another 37 air terminals are planned for the same period, together with 19 hotels and a large number of servicing and administrative buildings. The main airports are all to be equipped with ILS suitable for automatic or semi-automatic landing.

After the war the main Moscow airport was Vnukovo which had a small old-fashioned and generally inadequate terminal building. Bykovo was also in use. When air transport began its rapid growth in the late 1950s, work began in providing Moscow with more adequate airports. The Vnukovo terminal was enlarged and a new terminal constructed, these now being known as Vnukovo 1 and Vnukovo 2. Vnukovo 1 handles services to the Caucasus, Crimea and Ukraine, while Vnukovo 2 handles mainly Tu-124 services.

Bykovo is the headquarters of the Moscow Directorate of Aeroflot and is used by services to points in the Central Russian District and by cargo services.

In June 1961 Sheremetyevo Airport was opened. This has separate international and domestic terminals, the former being used by Aeroflot international services and all foreign carriers serving Moscow, and the latter Aeroflot services to the western part of the USSR and Polar services. The airport also handles An-12 cargo operations. Its latest terminal has a large circular cantilever roof similar to that of the Pan American World

Austerity air transport. Vnukovo Airport, Moscow, soon after the return of peace. Camouflaged Aeroflot Li-2s on the apron in front of the terminal building.

Classical Soviet airports. Top, Omsk in Western Siberia; centre, Alma Ata in Kazakh SSR; and bottom, Vil'nyus in Lithuanian SSR.

Airways' terminal at New York's John F. Kennedy International Airport. There is a single concrete runway, 07 - 25, measuring 3,500 m (11,483 ft) by 80 m (262 ft) and having approach lighting at each end. The inner sections of the lights consist of a double row on the left and a single row on the right.

Most recent of the Moscow airports is Domodedovo, opened in 1965, which handles the long-distance domestic services. The three-storey terminal is of concrete, light metal and glass construction, has two 200 m (656 ft) traffic fingers and can handle 3,000 passengers an hour.

Moscow also has a modern city air terminal and a heliport with regular services to the airports.

Baku, Gor'kiy, Kiev, Mineral'nyye Vody, Novosibirsk, Odessa, Ufa and Yerevan are among the cities with new airports or airport terminals, but many airports have outdated, if picturesque, terminal buildings

Il-18s at Moscow's Domodedovo Airport in 1964 before the new airport was officially opened in 1965. Behind the aircraft are the main terminal and control tower and to the right one of the traffic fingers. The nearest Il-18s are SSSR-75772 and SSSR-75770.

ranging from simple wooden structures of church-like appearance to highly unsuitable wedding-cake edifices. Many of these terminals possess considerable charm and it is to be hoped that when they are replaced by modern structures some of the finer examples are preserved. A survey of Soviet airport architecture would make a fascinating book.

Noise abatement procedures are in force at at least some Soviet airports, and complaints about noise appear to have to be taken as seriously as elsewhere.

Air traffic control procedures are similar to those in other parts of the world. The calmness of controllers is emphasized by the story of an airline pilot who was number two to land. He saw the aircraft in front of him

A modern airport terminal, at Dushanbe in Tadzhik SSR.

crash on the approach and was immediately told by the controller 'You are now turn one'.

At airports with ILS, Aeroflot's jet limits are 700 m (2,296 ft) runway visual range and no cloud limitation for take-off, and 1,000 m (3,280 ft) RVR and 100 m (328 ft) cloud base for landing. Thirty per cent is added to these figures if ILS is not available. These limits are much more conservative than in most countries. New aircraft are being equipped for automatic approach down to 30–40 m (98–131 ft).

Conclusion

The story of Aeroflot's development and the aircraft produced to meet the airline's requirements is a fascinating one. Although much development and modernization is still needed, particularly in airports and equipment, the story is one of which the Soviet civil aviation authorities can be proud and for this reason it is to be regretted that those responsible have seen fit to be so secretive about their achievements.

Mi-10 in the United States. Air-to-air view of Petroleum Helicopters' Mi-10 flying over Louisiana. Assembled in the US, this helicopter made its first flight in November 1967, is painted orange-yellow and black and bears the temporary registration Nl6556. It is to be used on oil operations in South America.

Antonov An-2

The Antonov An-2 single-engine transport biplane is absolutely unique and must be regarded as one of the world's truly great aeroplanes. It is unique because it is the only biplane transport to be designed and put into production since the 1939–45 war, and after 20 years' service it is still in series production. It is a great aeroplane because of the variety of tasks it has undertaken and because of the enormous contribution it has made to Soviet transport and agriculture. As a rugged hard-working aeroplane it should be compared with the Noorduyn Norseman and the de Havilland Canada Beaver. Judged by the number of variants and the geographical extent of its work, it must be put into the same category as the Douglas DC-3 and its Soviet counterpart the Lisunov Li-2.

The An-2 was designed by Oleg Antonov's bureau to meet a specification of the Soviet Ministry of Agriculture and Forestry. It was originally given the designation SKh-1 (Selskokhozyaistvennyi-1), and at one time seems to have borne the type name Kolkhoznik (Collective Farmer). The prototype, powered by a 630 hp Shvetsov ASh-21 engine, first flew in 1947.

Since that time the An-2 has been built in 17 or 18 versions for some 30 different duties. Up to 1960 the Soviet Union had built more than 5,000, and others had been produced in China and at Dresden in the German Democratic Republic. In 1960 Poland began producing An-2s and both

A standard Aeroflot An-2P flying in the Krasnoyarsk area of Siberia.

An orange-painted Polar Aviation An-2, SSSR-N542, on skis. The airscrew is the V509A-D7 type with curved blades.

the USSR and Poland were still building them at the end of 1967. Civil operation of the type began in March 1948.

In layout the An-2 is a large single-engine strut and wire-braced biplane of unequal span, with single fin and rudder and non-retractable wheel, float or brake-equipped ski undercarriage.

The wings are two-spar metal structures with parallel chord and have dihedral. Aft of the front spars they are fabric covered. The upper wing has full-span electrically-operated automatic slots, and the full span of the trailing edge is occupied by drooping slotted ailerons and slotted flaps. The lower wing has full-span slotted flaps. The wings are connected by I struts and wire braced. The fuselage is a stressed-skin all-metal semi-monocoque structure with circular section forward, rectangular section in the cabin area and oval section aft. The tail surfaces are metal structures, the fin being built integral with the fuselage and the strut-braced tailplane being attached near the base of the fin. The tailplane is believed to be fabric covered aft of the front spar. The undercarriage of the landplane comprises divided-type main unit, with long-stroke oleos and low-pressure tyres, and a self-centring tailwheel. Tyre pressure can be changed in flight.

The standard engine is the 1,000 hp ASh-62IR nine-cylinder air-cooled radial which drives either a four-blade V509A-D7 constant-speed metal airscrew with curved blades or a four-blade V509-D9 with straight blades and square tips.

The pilots' cabin, in most versions, has side-by-side seats for two crew, and the canopy has balcony-type side windows which provide very good downward view.

The main variants of the An-2 are: An-2P, the basic general-purpose transport type; An-2S, agricultural version of the An-2P; An-2V, shallow-draught twin-float seaplane; An-2L, water-bomber version of the An-2V; An-2ZA, high-altitude meteorological research aircraft; and the considerably modified An-2M agricultural aircraft. A total of 17 versions are available for export but the individual designations are not known.

The An-2P is in very widescale use and this is the version used by Aeroflot on hundreds of local services. It is known that as late as 1967 An-2s were in service with all Aeroflot Directorates except Magadan, Moscow Transport, Polar and TUMVL (International), although it is possible that they were actually in service with Polar Directorate. A very high proportion of all these An-2s were of the An-2P version and it has been reported that more than 2,000 have been used on local services. Aeroflot has stated that thousands are in use and that on 15 May, 1967, one of them carried the 100 millionth passenger to be flown by its An-2s.

The An-2 has a cabin measuring 4·1 m (13 ft 5½ in) in length, 1·6 m (5 ft 3 in) in width and 1·8 m (5 ft 11 in) in height, and the entrance door, on the port side, measures 1·46 m by 1·53 m (4 ft 9¼ in by 5 ft). When used for passenger carriage the An-2P can have four seats on the starboard side and three on the port with a central aisle, or eight passengers in four double seats with a side aisle, or 10 passengers carried on tip-up inward-facing seats attached to the side walls. The cabin is heated and ventilated and there is a lavatory at the rear. As a cargo aircraft the An-2P can carry 1,240 kg (2,733 lb) of payload, cabin volume is 12 cu m (423·78 cu ft) and the floor is stressed to carry 1,000 kg/sq m (204·8 lb/sq ft).

Apart from the very large number of Aeroflot An-2Ps, the type has been supplied to Air Guinée, Air Mali, CAAC (China), ČSA, Deutsche Lufthansa (DDR) which later became Interflug, Mongolian Airlines, Royal Nepal Airlines and Tabso. The CAAC and Nepali aircraft were built in China where the type was put into production in 1957 as the Fongshu or Harvester 2.

The second most important version of the An-2 was the agricultural An-2S. This version has a longer-stroke undercarriage and a hopper capable of carrying 1,400 litres (307 Imp. gal) of liquid chemicals for spraying or 1,200 kg (2,645 lb) of dry chemicals for dusting. For spraying there

One of Dosaaf's An-2s. With the door removed, these are used for parachute training.
(*Courtesy William Green.*)

The first Chinese-built An-2, in December 1957. The Chinese aircraft bore the type name Fongshu 2 (Harvester 2).

is a spray-bar extending the full span of the lower wing, and the system has a propeller-driven pump beneath the fuselage which produces a spray rate of 6·5–18 litres/sec (85–237 Imp. gal/min). A tunnel-type distributor beneath the fuselage is used for dusting and seeding. The hopper is filled through the top of the fuselage. Agricultural operating speeds are 155–165 km/h (83·63–89·03 kt) (96·3–102·51 mph).

The importance of agricultural aviation to the Soviet Union cannot be overstressed. In 1966 spraying and dusting of agricultural lands and forests covered an area of 63,200,000 hectares (about 158 mn acres) and the planned coverage for 1967 was 70 mn hectares (175 mn acres). By 1970 it is planned that no less than 115 mn hectares (287·5 mn acres) shall be treated. Although helicopters are used for this work, for many years the biggest share has been done by the An-2S.

In 1954–55 the An-2V was developed under the design bureau designation An-4. The An-2V has two shallow-draught metal floats which each measure 7·5 m (24 ft 7¼ in) in length and are fitted with pneumatically-operated water rudders. This version can operate in water as shallow as 80 cm (2 ft 7½ in). These shallow-draught floats can be fitted to any version of the An-2 and conversion from wheels to floats can be made by four men in 20 hr; the change back to wheels can be made in 8 hr. Payload of the An-2V is 920 kg (2,028 lb). This version is known to be based at Petrozavodsk and serves lakeside villages in western Karelia.

1949 saw the appearance of a firefighting version of the An-2 and in 1964 the An-2L* water-bomber version of the An-2V was produced. The water-bomber can carry 1,260 litres (277 Imp. gal) of water in the floats and this is dropped from exits just forward of the float steps. Water is picked up in 5–7 sec while taxi-ing at 45–50 km/h (24·28–26·98 kt) (27·96–31·06 mph). Dropping is normally done at 150–160 km/h (80·94–86·34 kt) (93·21–99·42 mph) from a height of 10 m (32 ft) and the area covered is 12–14 m (39–46 ft) wide and 65–70 m (213–229 ft) long.

An-2s are widely used as ambulances and can carry six stretcher cases, or three stretcher cases, six sitting cases and a medical attendant. No

* Described in Soviet reports as An-2P.

special designation is known for the ambulance version. In the Soviet Union there are about 200 air ambulance stations and from these An-2s, Yak-12s and helicopters make about 100,000 flights a year carrying some 200,000 doctors and patients and 500 tons of medical supplies.

Apart from the specialized duties already described, An-2s are used for dropping parachutists, with up to 14 being carried; Polar work, when they are painted bright orange; photography and photographic survey; glider-towing; and geological and maritime exploration.

One rather strange version of the An-2 is the high-altitude meteorological research An-2ZA, which bears the design bureau designation An-6 and has an enclosed cockpit above the rear fuselage immediately ahead of the fin. The engine used is a turbo-supercharged ASh-62IR/TK which maintains 850 hp up to a height of 9,500 m (31,168 ft).

In addition to the wide range of An-2s used by Aeroflot and supplied to other airlines, numbers are used by the Soviet Air Force and by Dosaaf, the semi-military flying club organization. The United Arab Republic has adopted the An-2 for its Air Force, the Cuban National Agrarian Reform Institute (INRA) has some including CU-E802, and several Hungarian Air Force An-2s from the Dresden production line have gone on to the civil register and include HA-ANA, HA-ANB and HA-JUE. SP-ANP, SP-ANS, SP-ANT and SP-ANU are among those used for agricultural work in Poland. In Czechoslovakia ČSA's agricultural department, Agrolet, uses four An-2s, and OK-MYA, OK-MYB and OK-MYC are believed to be three of them. It has been stated that in 1950

One of Aeroflot's shallow-draught An-2V floatplanes. This example, SSSR-23812, was brownish-green with yellow spinner. The crest beneath the cockpit was light blue and yellow, and on the bow of the port float was the number 12105. (*Courtesy Aviation Magazine International.*)

One of the An-2Ps used by the East German Deutsche Lufthansa. These aircraft, which had rectangular cabin windows, passed to Interflug.

alone 100 An-2s were sold to Bulgaria, Czechoslovakia, East Germany, Mongolia, Rumania and Yugoslavia.

Although some of Aeroflot's An-2s are smartly painted, many are finished in olive drab with blue-grey undersurfaces and their registrations are applied with stencils. The An-2Ps used by Deutsche Lufthansa (DDR) and which passed to Interflug differ from other An-2s in having rectangular cabin windows in place of the normal circular type.

In 1964 the Soviet Union began quantity production of what might be called the first major-change An-2. This was the An-2M agricultural aircraft which incorporated the experience, in the USSR alone, of more than 3,000 mn km (1,864 mn miles) operation by the previous models. The new An-2M was designed for fertilizing, crop feeding, pest control, weed killing and the defoliating and desiccating of cotton plants. In carrying out this work the An-2M was intended to be 15–20 per cent more efficient than the An-2S and to cut costs by 22–27 per cent per hectare (2·471 acres).

The An-2M retains the same general appearance of earlier An-2s but has the undercarriage moved forward by 7 cm (2¾ in) to reduce the risk of nosing-over, and the tail unit has been completely redesigned with increased-area tailplane and with straight-edged fin and rudder. The airframe has been improved to increase its service life by having metal bonded and welded joints and improved corrosion protective coating. The control cabin has been redesigned for single-pilot operation, is air-conditioned and separated from the cargo hold by a pressure bulkhead. It is entered via the port side of the canopy which now folds upward to form a door, and there are two steps on the port side of the fuselage to provide access from the lower wing. The engine used is still the 1,000 hp ASh-62IR but the airscrew is a four-blade variable-pitch AV-2. Six tanks in the upper wing have a total fuel capacity of 1,200 litres (263 Imp. gal). The tailwheel can be locked to make for easier cross-wind take-offs and landings, and the mainwheel tyre pressure is 3 kg/sq cm (42·67 lb/sq in).

Improvements have been made in the spraying equipment and more uniform distribution has been achieved. The fibre-glass chemical tank has 1,960 litre (431 Imp. gal) capacity. The dispensing system is driven by the engine via a shaft and 50 hp gearbox. The spray-bar has been increased in length and projects well beyond the tips of the upper wing to which it is braced by V struts. For dry chemical spreading, beneath the fuselage there is a three-channel tunnel sprayer made of a chemical-resistant aluminium alloy. The chemicals are forced out by the airstream through the forward bifurcated intake. Output is controllable. Dry chemicals are loaded

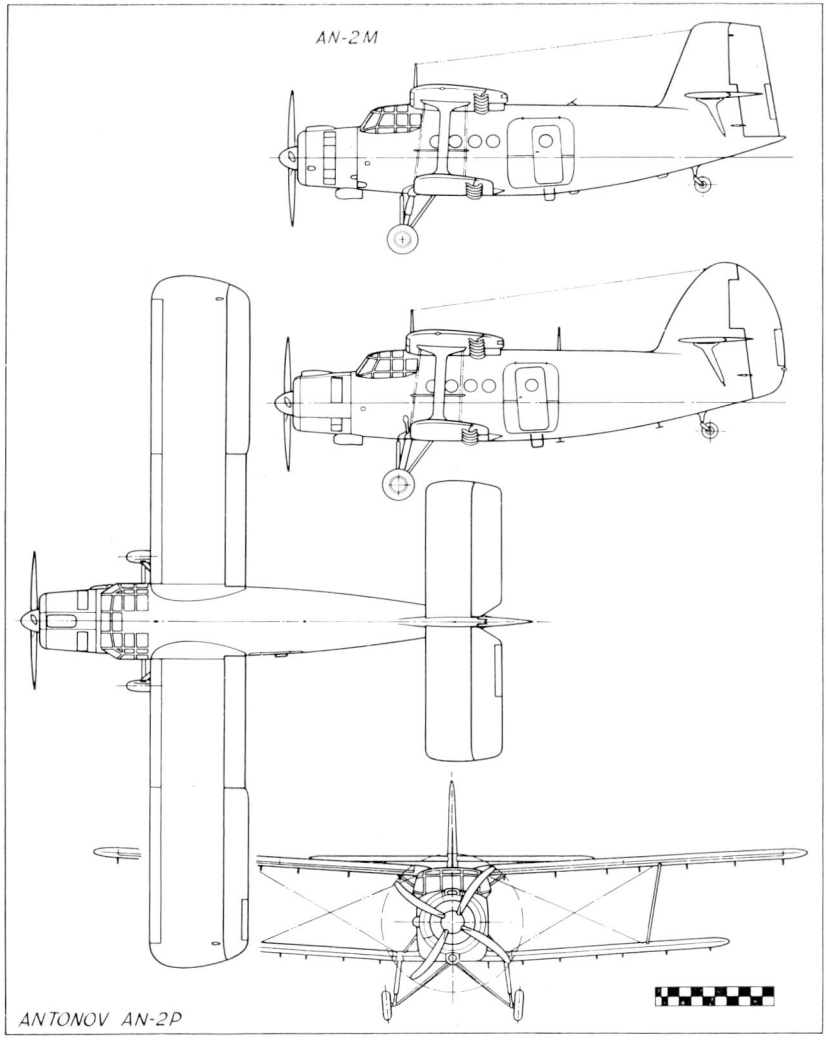

AN-2M

ANTONOV AN-2P

An Antonov line-up. On the left is the An-2M SSSR-05901 with dusting equipment. Next to it is the An-2M SSSR-05902 with spray-bar. At the end of the line is the production An-14 SSSR-81550.

through a hatch in the top of the fuselage and liquid loads are taken in through a connection under the centre section.

Cabin dimensions of the An-2M are the same as those of the earlier types. The door measures 1·65 m by 1·67 m (5 ft 5 in by 5 ft 5¾ in) and is hinged to open upward. Within the main door is an aft-opening passenger door in which there is a circular window. The An-2M can be used as a 12-passenger or cargo transport, can be fitted with dual control, and in the transport version there is a door in the pressure bulkhead to provide access to the control cabin.

Aeroflot An-2Ms have light grey fuselages with white tops, white wings and tail units. They normally have coloured spinners and a coloured horizontal band across the fin and rudder. Registrations are known in the SSSR-05900, 23000, 33000, 42600 and 62700 series.

Latest version of the An-2 is the An-2M which has a completely redesigned tail unit. The Aeroflot example seen here is fitted with dusting equipment.

The Polish An-2s are built by Wytwornia Sprzetu Komunikacyjnego-Mielec at Mielec and have been produced in the following versions: An-2R (Soviet An-2S), An-2P (as the standard Aeroflot aircraft), An-2W (Soviet An-2V), An-2ZA (as Soviet An-2ZA), An-2SN ambulance, An-2T (mixed passenger–cargo version, superseded in 1967 by a passenger variant designated An-2P) and An-2M (as Soviet An-2M).

An-2

Span 18·18 m (59 ft 7¾ in) upper, and 14·24 m (46 ft 8½ in) lower; length 12·4 m (40 ft 8 in); height 4·13 m (13 ft 6½ in) tail down, and 5·35 m (17 ft 6½ in) tail up; wing area 71·52 sq m (769·83 sq ft); aspect ratio 7·7 upper, and 7·25 lower; track 3·36 m (11 ft 0¼ in).

Empty weight 3,400 kg (7, 495 lb); fuel 900 kg (1,984 lb); payload 1,300 kg (2,866 lb); maximum take-off weight 5,500 kg (12,125 lb).

Maximum speed at 1,750 m (5,741 ft) 256 km/h (138·13 kt) (159·06 mph); cruising speed (transport version) 190 km/h (102·52 kt) (118·06 mph), (agricultural version) 160 km/h (86·34 kt) (99·42 mph); landing speed 84 km/h (45·32 kt) (52·19 mph); take-off run on concrete 150 m (492 ft), on earth surface 180 m (590 ft); landing run on concrete 170 m (557 ft), on earth surface 180 m (590 ft); service ceiling 4,500 m (14,763 ft); range at 1,000 m (3,280 ft) with 670 litre (147 Imp. gal) fuel reserve 845 km (455 n.miles).

Take-off and landing run of agricultural version with limited fuel 80–100 m (262–328 ft).

An-2M

Span 18·18 m (59 ft 7¾ in) upper, and 14·24 m (46 ft 8½ in) lower; length, tail down, 12·95 m (42 ft 5¾ in); height, tail down, 4·2 m (13 ft 9¼ in); wing area 71·1 sq m (765·31 sq ft); track 3·36 m (11 ft 0¼ in).

Empty weight 3,430 kg (7, 561 lb); fuel 900 kg (1,984 lb); maximum chemical payload 1,500 kg (3,306 lb); maximum take-off weight 5,500 kg (12,125 lb).

Maximum speed at 1,750 m (5,741 ft) 253 km/h (136·51 kt) (157·2 mph); economic cruising speed 200 km/h (107·92 kt) (124·27 mph); spraying speed 150–160 km/h (80·94–86·34 kt) (93·21–99·42 mph); take-off speed 85–90 km/h (45·86–48·56 kt) (52·81–55·92 mph); landing speed after chemical discharge 75 km/h (40·47 kt) (46·6 mph); maximum take-off run 200 m (656 ft); landing run after chemical discharge 100 m (328 ft); maximum rate of climb at nominal power (820 hp) with flaps retracted 2·8 m/sec (550 ft/min) clean aircraft, and 2 m/sec (393 ft/min) with spray-bar; service ceiling 4,350 m (14,271 ft); maximum fuel range 905 km (488 n.miles).

Maximum output. Granular chemicals 60 kg/sec (132 lb/sec), powder chemicals 37 kg/sec (81 lb/sec) and liquid chemicals 28 litres/sec (6·1 Imp. gal/sec). Maximum spray width 30–31 m (98–101 ft) during fertilizing and 38–42 m (124–137 ft) during liquid spraying. Performance figures are for maximum weight take-off in ISA.

An-2V

Payload 920 kg (2,028 lb); maximum take-off weight 5,250 kg (11,574 lb). Maximum speed at sea level 247 km/h (133·27 kt) (153·48 mph); minimum speed 87 km/h (46·94 kt) (54·05 mph); take-off run with 30 deg flap 80–100 m (262–328 ft); alighting run with 40 deg flap 60–80 m (196–262 ft); ceiling 5,000 m (16,404 ft); maximum payload range 1,480 km (798 n.miles).

A prototype An-8 military transport, first seen in June 1956.

Antonov An-8, An-10 and An-12

On the occasion of the 1956 Soviet Aviation Day display at Tushino a hitherto completely unknown twin-engine military transport aircraft was flown past the crowds. This was the Antonov An-8 on which design work must have begun at least by 1954. The first of five prototypes made its maiden flight in the autumn of 1955. A total of 100 production An-8s were built for the Soviet Air Force, which is reported to have given the type the designation An-4. Although the production run was not all that small, surprisingly little is known about the An-8. It was a high-wing monoplane with high aspect ratio wing which had a straight trailing edge and sweptback leading edge. Flaps and ailerons extended the entire length of the trailing edge. The fuselage was of square section and upswept aft of the hold. Under the upswept rear was a large loading door, and the extreme rear of the fuselage housed a gun turret. There were eight or nine circular windows in the port side of the hold of the production aircraft and probably the same number in the starboard side.

The An-8's main undercarriage consisted of two tandem pairs of wheels on each unit and these retracted into large blisters on the sides of the fuselage. There was evidence of considerable fuselage reinforcement between the undercarriage pick-up points and the wing spars.

The engines were two Kuznetsov NK-2M airscrew-turbines each of 5,100 hp. Four-blade airscrews were used.

The An-8 had a nose position for the navigator, and chin radar was fitted. Production of the An-8 was relinquished in favour of the more advanced An-10, but An-8s remained in service for several years and some may still have been in use in 1967. Two or three were used experimentally on cargo operations by Aeroflot but the airline did not adopt the type.

In November 1955, before the An-8 was known, Antonov's design bureau had begun work on planning the bigger four-engine An-10. Construction of the first prototype was undertaken in record time and on 7 March, 1957, test pilot Ya. I. Vernikov took the An-10 on its first flight. Registered SSSR-U1957 and bearing the soon abandoned type name Ukraina (Ukraine), the prototype An-10 was shown at Vnukovo in July 1957 along with the prototype Il-18 Moskva (Moscow), the Tu-104A and the Tu-110.

The An-10 was designed for high-density traffic routes, to have high cruising speed but very good take-off and landing performance with ability to operate from unpaved surfaces. The prototype had accommodation for 84 passengers in three cabins.

In layout the An-10 was a high-wing cantilever monoplane, with circular-section pressurized fuselage, single fin and rudder and multi-wheel undercarriage. The four engines in the prototype were 4,000 ehp Kuznetsov NK-4 turbines each driving a four-blade airscrew.

It is believed that a number of the early An-10s were powered by Kuznetsov engines but the 4,000 ehp Ivchenko AI-20 was finally chosen as the standard engine. When production An-10s appeared it was seen that a number of major modifications had been made, most of them because of stability problems encountered during the test programme. The outboard sections of the wing were given marked anhedral, the engines were lower-slung and the flight deck windows had been changed. Between its roll-out and its appearance at Vnukovo in July 1957, the prototype had received a

Paratroops boarding a Soviet Air Force An-8. (*Courtesy William Green.*)

The prototype An-10 Ukraina (Ukraine) SSSR-U1957 at Vnukovo Airport, Moscow, in July 1957. The dorsal fin had been deepened since the aircraft was rolled-out.

much deeper dorsal fin, and it is possible that the vertical fin had been heightened. Changes in tail design were to continue for some time and the first major change was the fitting of vertical fins at the ends of the tailplane.

It is normal Soviet practice to put new transport aircraft into service as freighters before they begin passenger service, and it is known that cargo operations were undertaken by An-10s. Two cargo flights were made on 27 May, 1959, the routes not being known, but in August 1959 it was stated that for about two months An-10s had been operating cargo services within the Ukraine, to Western Siberia and in the Arctic regions.

On 27 May, 1959, a technical flight with passengers was made from Kiev to Moscow, Tbilisi, Adler and Khar'kov before returning to Kiev. On 21 July, 1959, an An-10 carried passengers from Kiev to Moscow and this was almost certainly a positioning flight by SSSR-11158 which inaugurated regular An-10 service on 22 July when it flew from Moscow to Simferopol with 85 passengers. SSSR-11158 left the Vnukovo ramp at 09·50 Moscow time, was airborne at 09·52 and, after cruising at 7,000 m

The 46-seat main cabin of the prototype An-10.

Early production An-10, SSSR-11141, at Vnukovo. In the background are Tu-104s
(*Flight International.*)

(22,965 ft), landed at Simferopol at 12·05. Kiev–Simferopol An-10 services started the same day and the first operations of the type were all by Aeroflot's Ukrainian Directorate.

Soon after the appearance of the An-10, Antonov produced a slightly modified version known as the An-10A. This was 2 m (6 ft 6¾ in) longer and had accommodation for 100 passengers. The An-10A is known to have gone into service with Aeroflot in February 1960, and the first regular operation of the type was probably between Moscow and Rostov on which route the An-10A was introduced on 10 February.

In 1967 six Aeroflot Directorates were operating An-10s and four were using An-10As. The An-10s were working 26 Federal routes and the An-10As 44 Federal routes in the summer of 1966, at which time four Directorates were also known to be using An-10s on local services. Aeroflot stated that in 1966 An-10s and An-10As were operating on more than 90 routes and carried 4,200,000 passengers. Between July 1959 and July 1967 the An-10 type made 100,000 flights and carried 6 mn passengers and 400,000 tons of cargo and mail within the Ukraine. By mid-1967, on all

The production An-10 SSSR-11158 leaving Vnukovo Airport, Moscow, possibly on 22 July, 1959, when it operated the first regular scheduled An-10 service, from Moscow to Simferopol. This production aircraft differs from the prototype in having anhedral on the outer wing, lower-slung engines and end-plate fins.

Aeroflot's An-10A SSSR-11220, with twin vertical fins.

operations, these aircraft had carried 12 mn passengers and 800,000 tons of cargo and mail. The 10 millionth passenger carried by An-10s and An-10As was carried aboard SSSR-11137 on 30 May, 1966.

All the An-10s have been worked as 85-passenger aircraft and for several years the An-10As had 100 seats, but more recently the Moldavian and Ukrainian Directorates' An-10As have had accommodation for 110 passengers. Some years ago it was reported that Antonov was working on a developed An-10A which would be 6 m (19 ft 8 in) longer and have accommodation for up to 132 passengers. This type was known as the An-16 and Antonov said it would probably go into service as the An-10B (or V). Aeroflot has mentioned a 132-seat tourist version but there is no evidence that this has ever gone into service.

After the An-10As had been in service for some time yet another tail modification was made. The twin fins were removed and the original single ventral fin was replaced by two angled-out fins under the rear fuselage. This modification does not appear to have been made to the An-10s.

The number of An-10s and An-10As is not known but it is possible that Aeroflot took delivery of as many as 300. The earliest known registration for a production aircraft is SSSR-11129 and the highest identified in this series is SSSR-11229. There is also the out-of-series aircraft SSSR-34385.

The An-10A SSSR-11185 with angled-out ventral fins. Standard An-10 and An-10A markings are in red.

Much less is known about the An-10 and An-10A than about many other Soviet transports because no attempt has been made to export them, but even so a reasonably clear picture of its structure and interior layout can be given.

The wing tapers in chord and thickness but all the taper is on the leading edge. It is a two-spar all-metal structure built in five sections. The centre section extends only for a short way on each side of the fuselage,

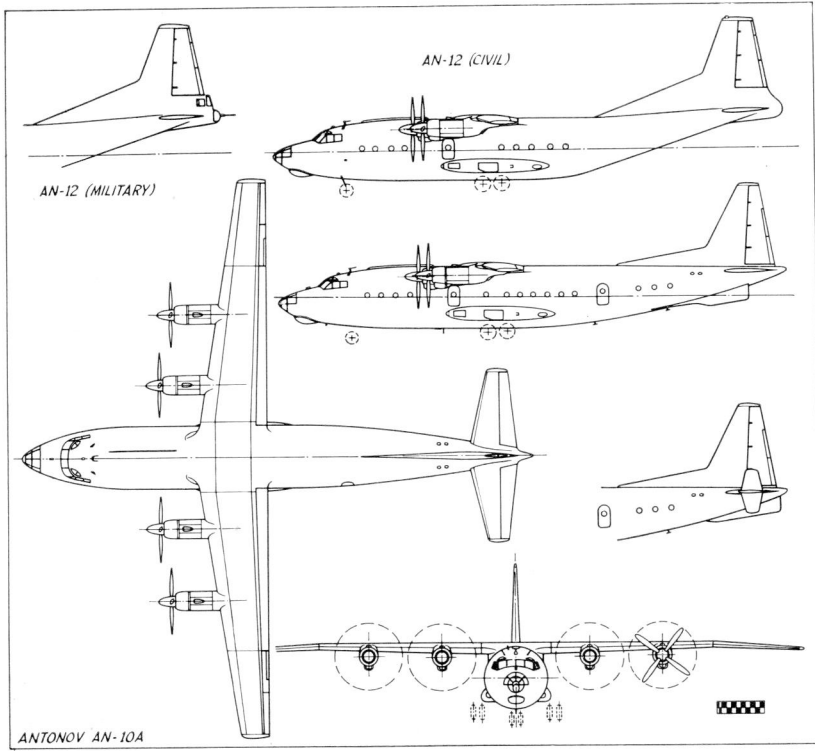

the inner sections carry the engines and the outer sections have anhedral. Manually-operated aerodynamically-balanced ailerons occupy the full span of the outer wing panels and incorporate tabs. The inner wing sections carry hydraulically-operated full-span double-slotted Fowler-type flaps. Electro-thermal leading edge deicing is provided and the wings contain 22 flexible bag tanks with a total fuel capacity of 11,040 kg (24,339 lb).

The fuselage is a circular-section semi-monocoque stressed-skin structure with the underside swept up quite markedly to the tail cone. Pressure differential is 0·5 kg/sq cm (7·1 lb/sq in).

The tail unit is of all-metal construction and control surfaces are aerodynamically balanced and manually operated. Rudder and elevators have electrically-operated trim tabs and all leading edges have electro-thermal ice protection.

The main undercarriage units each have a bogy with two pairs of wheels in tandem and these are hydraulically-retracted upward and inward to be housed in large fuselage blisters. Disc brakes are fitted and tyre pressure is 5·6–6·7 kg/sq cm (79·65–95·29 lb/sq in). Twin steerable nosewheels retract into the forward fuselage and the doors are open only while the wheels are in transit. There is a retractable tail bumper. The undercarriage-housing blisters also contain the cabin pressurization system. It has been reported that An-10As operate on skis in the Arctic regions.

The four Ivchenko AI-20 engines now in standard use are of the AI-20K version and develop 4,000 ehp for take-off. They are single-spool engines with 10-stage axial-flow compressor and three-stage turbine. The airscrews are AV-68 reversible-pitch four-blade units of 4·5 m (14 ft 9 in) diameter.

Crew comprises two pilots, engineer and radio operator on the main flight deck and a navigator forward at a lower level.

The exact layout of the 85-seat An-10 is not known but is probably the same as that of the prototype except for the addition of one seat. In the prototype immediately aft of the flight deck was the forward cabin with four circular windows on each side. This cabin had 25 seats—the three back rows being in triple units on each side and the front full row being arranged with three seats to starboard and two to port. Right forward on the port side were two aft-facing seats, and the equivalent position on the starboard side was occupied by a lavatory. Aft of this front cabin was a

Standard military An-12. This one, with Himalayan background, was one of a number supplied to the Indian Air Force. (*Indian Air Force.*)

Soviet Air Force An-12s on the runway at Domodedovo during the big display in July 1967. The ramp of the nearest aircraft is being closed after the unloading of a tank.

wardrobe to starboard and to port a baggage hold of 5 cu m (176·57 cu ft) capacity. Then came the forward entrance door which was on the port side above the undercarriage blister and, on the starboard side, the galley. Aft of this was the main cabin with seats for 46 passengers. There were six rows of six-across and, aft on the starboard side opposite the rear entrance, two rows of triple units. Right forward were two double seats and beside these against the fuselage walls were seats for children. These forward seats could be curtained off so that they were separated from the main cabin — complete privacy was ensured because there were no cabin windows in line with these seats! There was a window beside each of the other seat rows, making six on the port side and eight on the starboard. Aft of the rear entrance was another lavatory, and a cabin with 13 seats arranged in three pairs on the starboard side, two pairs to port and a row of three against the rear pressure bulkhead. This aft cabin had three windows in each side. Luggage racks ran the length of both sides of each cabin and across the back bulkhead of the aft cabin. There were framed pictures on the bulkheads and between the windows and luggage racks.

In the 100-seat An-10As, which have a total cabin volume of 222 cu m (7,839·8 cu ft), the forward cabin has seats for 26, with four rows of six-abreast and two aft-facing seats on the port side against the front bulkhead. There is still a lavatory on the starboard side. Then comes the wardrobe, baggage hold, entrance door and galley as on the An-10, and behind this are two small cabins each with five seats—three forward-facing and two aft-facing. The main cabin has 42 seats arranged in seven rows of six-abreast. Behind the main cabin are the rear entrance and two lavatories and then a 16-seat cabin with two triple and two double seats on the starboard side and three pairs of seats on the port side. The rear pressure bulkhead has been moved further aft and this has made room for a small cabin with three inward-facing seats on each side. In both the An-10 and An-10A the cabin floor slopes up towards the rear in the area aft of the wing. It was on An-10As on the Moscow–Kiev route that Aeroflot began showing films in flight.

The An-10s and An-10As have formed a significant part of Aeroflot's fleet and large numbers are still in service, but over the next few years they will be replaced, mainly by Tu-154s.

Built at the same time as the original An-10 was the An-12 military cargo aircraft. The production An-12s are the same as the production An-10s but have a completely new rear fuselage which is sharply upswept and has a flat underside incorporating loading doors. The ventral fin was dispensed with and the dorsal fin deepened. Above the horizontal tail surfaces and immediately beneath the rudder is a tail gun position.

SSSR-11357, a Polar Aviation An-12. It appears to have the standard red paint scheme applied to an overall orange finish. In the background are another Polar Aviation An-12, SSSR-04343, and a Polar Il-14, SSSR-04177.

The An-12 became standard Soviet Air Force equipment for freight and vehicle transport and for carriage of paratroops. Numbers have been supplied to Algeria and to the Indian, Indonesian, Iraqi and United Arab Republic Air Forces.

Aeroflot had some An-12s with the armament removed, Ghana Airways had the all-white 9G-AAZ and Cubana had an An-12 registered CU-T827. At least two of the UAR aircraft have been operated with civil registrations, SU-AOI and SU-AOJ, and there was a Polish An-12 registered SP-LZA. One United Arab Air Force An-12 was used for flight testing the Egyptian Helwan E-300 turbojet which was installed in place of the port inner AI-20.

At the 1965 Paris Aero Show a modified and completely civil version of the An-12 was exhibited and flown. The actual aircraft was SSSR-11359 and this bore the designation An-12 on its nose and rudder, but later the

Ghana Airways' minimum-change An-12 9G-AAZ at Accra. This aircraft retained the tail gunner's position although the armament had been removed, and was white overall with red-outlined bright blue fuselage stripe and the red, yellow and green national colours and black star on its rudder. (*Ghana Airways*.)

Cubana's An-12 CU-T827 which also retained the rear gunner's position. This aircraft was lost on the approach to Mexico City on 9 February, 1967.

Aeroflot An-12 SSSR-11366 at Orly Airport, Paris, on the occasion of the inauguration of the joint Air France–Aeroflot Paris–Moscow cargo service in February 1966. SSSR-11366 retains the tail gun position but has the windows blanked-off. The large size of the under-fuselage freight door can be seen. (*Air France*.)

SSSR-11359, a purely civil An-12 shown at the 1965 Paris Aero Show. The upper fuselage and fin were white, fuselage markings and rudder bright blue. (*R. A. Cole.*)

same aircraft was exhibited in the USSR as the An-12B and this may be the correct designation for this purely civil model.

The civil An-12 has a redesigned tail, from which the turret has been deleted, and a redesigned interior. The main hold measures 13·5 m (44 ft $3\frac{1}{2}$ in) in length, 3 m (9 ft $10\frac{1}{4}$ in) in width and 2·4 m (7 ft $10\frac{1}{2}$ in) in height. To some extent the floor area is obstructed by the main frames beneath the spars and by the inward-opening forward entrance door. The rear part of the floor slopes down to the under-fuselage doors. The forward part of the door is divided into two sections which fold upward against the cabin walls, and the rear part of the door folds upward within the fuselage.

The redesigned rear fuselage of the An-12 SSSR-11359. Although the designation Antonov 12 appears on the base of the rudder, the true designation is believed to be An-12B. (*John Stroud.*)

The cargo floor is stressed for loads of 1,500 kg/sq m (307 lb/sq ft), and for loading there is an overhead hoist capable of carrying 2,300 kg (5,070 lb). There are lashing points in the floor. Between the main hold and the flight deck is a pressurized cabin with 14 seats. The aircraft seen at Paris had curtains at the main hold windows. Tip-up seats can be provided along the side walls. The rear doors can be opened in flight and loads can be para-dropped. A detachable ramp can be used to embark vehicles.

Hold, loading doors and rear ramp of the An-12B SSSR-11359. The overhead hoist can be seen at the rear of the cabin. On the left is a Tu-124. (*Flight International*.)

In 1967 the An-12 was in service with Aeroflot's Eastern Siberia, Magadan, Polar, TUMVL (International), Ukrainian, Ural and Volga Directorates, and in 1966 it was known to be operating scheduled cargo services over at least 11 Federal routes. On 3 February, 1966, SSSR-11366 opened a joint Aeroflot–Air France Moscow–Riga–Paris cargo service, and at about the same time An-12s began operating a regular cargo service between Moscow and Djakarta via Tashkent, Karachi, Colombo and Medan. It is thought that early in 1967 An-12s began operating cargo services from Moscow to Magadan and Petropavlovsk-Kamchatskiy.

In December 1961 one of the demilitarized An-12s, SSSR-04366, made a special flight from Moscow to Mirnyy in Antarctica in 48 hr 27 min. This was a Polar Aviation aircraft with a black and white penguin painted on its orange tail; another Polar An-12 was SSSR-04364. There were a number of An-12s with out-of-series registrations but the normal Aeroflot aircraft are registered in the SSSR-11300 series, 13 being known between SSSR-11340 and SSSR-11397. An Aeroflot An-12 which visited Heathrow late in 1967 bore the registration SSSR-11031 and the c/n 7345003. This registration is earlier than any known for production An-10s and An-12s, but the c/n suggests that the aircraft was built in 1967. It is possible that this aircraft was an early An-10, perhaps the third, built in 1957 and recently converted to An-12 standard, or that the c/n refers to the date of conversion. To add to the confusion, the manufacturer's plate bears the designation An-10.

The pilots' positions and instrument panel of the An-12B SSSR-11359. Two fans can be seen suspended from the roof. (*Flight International.*)

The number of An-10s and An-12s is not known but there is some evidence that the total exceeds 500. Where constructor's numbers are known their sequence does not tie up with civil registrations. It would seem that An-10s and An-12s came from the same production lines and used a common group of c/ns; the civil An-12s may very well be conversions from military aircraft or An-10s.

The latest known development of the An-12 has been the fitting of SSSR-11381 (c/n 402807) with a new type of ski undercarriage. These skis are of broad beam and have shallow vee bottoms. The skis are heated and also incorporate a braking system. Trials took place in 1966–67 in the Kolyma River area of north-east Siberia, and it is understood that this type of ski will now be standard equipment for Arctic and Antarctic operation.

An-8

Span 30 m (98 ft 5¼ in); length 26 m (85 ft 3½ in). Empty weight 21,000 kg (46,297 lb); fuel 9,100 kg (20,062 lb); payload 9,000 kg (19,841 lb); normal take-off weight 35,000 kg (77,162 lb); maximum take-off weight 40,000 kg (88,185 lb). Maximum speed 610 km/h (329·16 kt) (379·04 mph); cruising speed 480 km/h (259·01 kt) (298·26 mph); range 2,815 km (1,519 n.miles).

All An-8 data is approximate and should be used with caution.

An-10A

Span 38 m (124 ft 8 in); length 34 m (111 ft 6½ in); height 9·8 m (32 ft 1¾ in); wing area 121·73 sq m (1,310·28 sq ft); track 5·42 m (17 ft 9¼ in); wheelbase 10·82 m (35 ft 6 in).

Empty weight not known. Maximum fuel 11,040 kg (24,339 lb); maximum payload 14,500 kg (31,967 lb); maximum take-off weight 54,000 kg (119,050 lb).

Maximum speed 715 km/h (385·81 kt) (444·28 mph); cruising speed 600–660 km/h (323·76–356·14 kt) (372·82–410·11 mph); economic cruising speed at 10,000 m (32,808 ft) 630 km/h (339·95 kt) (391·46 mph); take-off speed 190–210 km/h (102·52–113·32 kt) (118·06–130·49 mph); landing speed 163 km/h (87·95 kt) (101·28 mph); take-off run 700–800 m (2,296–2,624 ft); landing run 550–650 m (1,804–2,132 ft); service ceiling 11,000 m (36,089 ft); maximum payload range with 1 hr fuel reserve 1,220 km (658 n.miles); maximum fuel range with 8,440 kg (18,606 lb) payload and no reserves 4,075 km (2,198 n.miles); maximum fuel range with 1 hr fuel reserve 3,050 km (1,645 n.miles).

The original An-10 had a maximum take-off weight of 51,000 kg (112,436 lb) and a maximum payload of 13,000 kg (28,660 lb). Range with 12,000 kg (26,455 lb) payload was 2,000 km (1,079 n.miles), with 10,000 kg (22,046 lb) payload 3,000 km (1,618 n.miles) and with 8,200 kg (18,077 lb) payload 3,500 km (1,888 n.miles).

An-12B

Span 38 m (124 ft 8 in); length 33·1 m (108 ft 7¼ in); height 10·53 m (34 ft 6½ in); wing area 121·73 sq m (1,310·28 sq ft).

Empty weight not known. Maximum fuel 14,620 kg (32,231 lb); maximum payload 20,000 kg (44,092 lb); maximum take-off weight 61,000 kg (134,482 lb).

Maximum cruising speed 640 km/h (345·34 kt) (397·68 mph); economic cruising speed 580 km/h (312·97 kt) (360·39 mph); take-off run at 54,000 kg (119,050 lb) 850 m (2,788 ft); landing run 860 m (2,821 ft); service ceiling 10,200 m (33,464 ft); range with 10,000 kg (22,046 lb) payload and 1 hr fuel reserve 3,400 km (1,834 n.miles); maximum range 4,000 km (2,158 n.miles).

The An-12B, SSSR-11359, flying past at the 1965 Paris Aero Show at Le Bourget.
(*Aviation Magazine International.*)

Known dates for introductions and technical proving flights:

An-10 and An-10A

1959 May 27	Two cargo flights made, routes not known (An-10)
1959 May 27	First technical flight with passengers, Kiev–Moscow–Tbilisi–Adler–Khar'kov–Kiev (An-10)
1959 July 21	Kiev–Moscow (inaugural flight with passengers, believed positioning flight by An-10 SSSR-11158)
1959 July 22	Moscow–Simferopol (first regular service, by An-10 SSSR-11158)
1959 July 22	Kiev–Simferopol (An-10)
1959 December 17	Start of Moscow–Washington (Andrews Air Force Base) flight by An-10A SSSR-11172 with Christmas trees for President of United States
1960 January 7	Kiev–Moscow–Delhi (special flight)
1960 February 3	Khabarovsk–Magadan (technical proving flight)
1960 February 10	Moscow–Rostov (An-10A)
1961 February 2	Moscow–Noril'sk
1961 February 21	Moscow–L'vov
1961 February 25	Khabarovsk–Magadan (An-10A)
1961 March 3	Moscow–Odessa
1961 March 12	Moscow–Donetsk, Moscow–Khar'kov and Moscow–Zaporozh'ye
1961 April 5	Irkutsk–Yakutsk
1961 May 15	Moscow–Adler/Sochi
1961 June 1	Moscow–Minsk
1961 June 22	Minsk–Leningrad and Minsk–L'vov
1961 November 17	Kiev–Minsk and Minsk–Baku
1961 December 12	Novosibirsk–Adler/Sochi (An-10A)
1962 May 15	Moscow–Kaliningrad
1962 May 17	Moscow–Ivano-Frankovsk (formerly Stanislav)
1963 January 11	Moscow–Blagoveshchensk
1963 February 5	Moscow–Ufa and Kuybyshev–Ufa
1963 May 15	Moscow–Kherson
1964 May 15	Moscow–Krivoy Rog, Moscow–Lugansk and Moscow–Volgograd
1964 December 1	Moscow–Vorkuta

An-12

1961 December	Moscow–Tashkent–Delhi–Djakarta–Sydney–Christchurch–Mirnyy (special flight by SSSR-04366)
1965 February 18	Cargo services linking Moscow, Leningrad, Minsk, Gor'kiy, Sverdlovsk, Chelyabinsk and Novosibirsk
1966 February 3	Moscow–Riga–Paris (by SSSR-11366, in association with Air France)
1966	Moscow–Tashkent–Karachi–Colombo–Medan–Djakarta

In the summer of 1967 An-12s were known to be operating the following daily cargo services:

Eastern Siberia Directorate. Irkutsk–Mirnyy,* Ust'-Kut–Mirnyy and Ust'-Kut–Yakutsk

* Operated thrice daily.

Magadan Directorate. Magadan–Yakutsk–Krasnoyarsk–Novosibirsk–Sverdlovsk–Moscow

Polar Directorate. Leningrad–Gor'kiy–Chelyabinsk–Novosibirsk, Moscow–Minsk–Gor'kiy–Sverdlovsk–Novosibirsk and Moscow–Sverdlovsk–Novosibirsk–Kemerovo–Irkutsk–Blagoveshchensk–Khabarovsk

Ukrainian Directorate. Khar'kov–Sverdlovsk–Novosibirsk and Zaporozh'ye–Sverdlovsk–Novosibirsk–Kemerovo–Sverdlovsk–Khar'kov–Zaporozh'ye

Volga Directorate. Gor'kiy–Chelyabinsk–Novosibirsk–Krasnoyarsk, Kazan'–Sverdlovsk–Novosibirsk, Kuybyshev–Novosibirsk–Krasnoyarsk and Kuybyshev–Ufa–Omsk–Novosibirsk

Details of operations by the Ural Directorate are not known.

Antonov An-14 Pchelka (Bee)

The Antonov An-14 Pchelka was designed to meet an Aeroflot specification for a general-purpose STOL aircraft and was most likely intended as a replacement for the single-engine Yak-12. The original requirement was for an aircraft with accommodation for pilot and three passengers, but subsequent demands called for five passenger seats and, later, seven. The first prototype, SSSR-L1958, made its first flight on 15 March, 1958, but the type did not enter service until 1965 when it was reported to have gone into regular operation on Aeroflot feeder services in the Vladivostok area.

SSSR-L1958, the first prototype An-14 Pchelka (Bee).

SSSR-L1958, the first prototype An-14 after modification.

It has been reported that the delay was due to Aeroflot's changing requirements, including the increase in seating, but this does not seem to be accurate because the September 1957 issue of *Grazhdanskaya Aviatsiya* contained four photographs of a full-size furnished mock-up of the An-14, and this certainly had accommodation for pilot and six or seven passengers.

In layout the An-14 is a strut-braced high-wing monoplane with twin fins and rudders and non-retractable nosewheel undercarriage. The fuselage is a metal semi-monocoque structure which tapers to a single boom aft.

Numerous modifications were undertaken between the first flight of the prototype and the start of quantity production. These changes involved the fuselage shape, the wing span and planform, the tail unit, underfuselage loading doors and engines.

The prototype, SSSR-L1958, was powered by two 260 hp Ivchenko AI-14R nine-cylinder air-cooled radial engines enclosed in close-fitting cowlings and driving three-blade airscrews which were fitted with large-diameter rather blunt spinners. The wing was of parallel chord, the tailplane was without dihedral and the fin leading edges were sweptback above and below the tailplane attachment points. This aircraft very closely resembled the full-size mock-up except that the mock-up had rectangular

SSSR-L1956, believed to be the second prototype An-14. It is seen here on skis and finished in a red and yellow colour scheme.

SSSR-L1053, one of the prototype An-14s, seen with spray-bars beneath the wing and attached to the wing bracing strut.

low aspect ratio fins and rudders without taper on the leading edges. The rear section of the forward fuselage could be slid aft and up to provide loading access to the cabin, and this entrance incorporated an outward-opening door divided into two sections horizontally along the centreline.

The second prototype appears to have been SSSR-L1956 which closely resembled SSSR-L1958 but had 300 hp Ivchenko AI-14RF engines with modified cowlings, two-blade square-tipped airscrews without spinners, and eyebrow cockpit windows. SSSR-L1956 is known to have been fitted with skis at one period. Another prototype was SSSR-L1053.

The next known An-14 was the silver, white and blue SSSR-L5860 which appears to have been the link between the early prototypes and the production aircraft. L5860 had increased span with taper outboard of the engines, a modified nose, parallel-chord inward-sloping fins and

An-14, SSSR-L1956, in agricultural configuration, showing loading door open. (*Aviaexport.*)

rudders, dihedral tailplane and new under-fuselage doors. Flap changes also seem to have been made and the size of the aileron brackets reduced.

In 1965 the An-14 entered service, and photographs of a production aircraft, SSSR-81550, showed further changes. The nose had been lengthened, slight changes had been made to the base of the fins, and the two-blade airscrews had been replaced by constant-speed three-blade units.

The wing comprises a parallel-chord centre section without dihedral and outer sections with 2 deg dihedral and with taper on the trailing edge. Details of the all-metal structure are not known. Double-slotted flaps, in two sections each side, occupy about 50 per cent of the span, and the ailerons are also slotted. Leading edge slats extend from outboard of the engines to the wing tips. A short stub-wing carries the main undercarriage units and the mainplane is braced to the stub-wing by a single tapered streamlined strut on each side.

The all-metal semi-monocoque fuselage is of rectangular section with rounded corners in the cabin area but immediately aft of the cabin is upswept on its underside to form a tail boom. The tail unit is a metal-skinned metal structure with inward-inclined parallel fins and rudders attached to the dihedral tailplane at their mid-points.

The undercarriage, designed for operation from rough and soft surfaces, comprises single-wheel main units and a single nosewheel. All wheels are the same size, the nosewheel is steerable and the main wheels have brakes. Skis can be fitted for winter operation, and it has been reported that twin floats can be fitted.

The Ivchenko AI-14RF nine-cylinder air-cooled radial engines develop 300 hp at take-off, have a rated power of 285 hp, a dry weight of 217 kg (478 lb), pneumatic starting, and drive 2·9 m (9 ft 6 in) diameter three-blade constant-speed feathering airscrews.

The cabin, excluding the flight deck, is 3·1 m (10 ft 2 in) in length, 1·6 m (5 ft 3 in) in height and 1·53 m (5 ft) in width. The standard passenger layout has three seats on each side and a seventh seat on the

SSSR-81550, a production An-14 with dihedral tailplane and inclined rectangular fins and rudders.

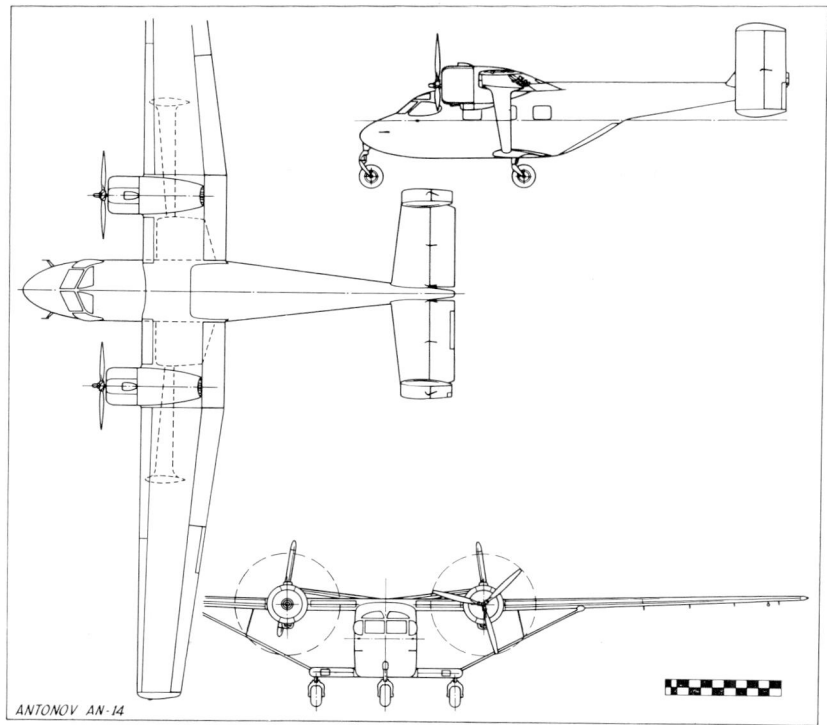
ANTONOV AN-14

starboard side of the flight deck, but two extra seats can be installed to bring the passenger total to nine. Seats are easily removable to provide unobstructed cargo space. A five-passenger executive version can be supplied and this has the four cabin seats arranged to face across fixed tables. Entrance to the An-14 is via the under-fuselage doors which provide an opening 1·9 m by 85 cm (6 ft $2\frac{3}{4}$ in by 2 ft $9\frac{1}{2}$ in). In the agricultural version it is believed that there is a forward door.

An ambulance version of the An-14 has a three-tier litter on each side and the medical attendant occupies the second flight deck seat.

For agricultural work the An-14 is fitted with a wide-span spray-bar and a 1,000 litre (220 Imp. gal) chemical tank.

The An-14 is simple in concept and construction, easy to maintain and is apparently viceless because it is claimed to be safe for the average pilot in spite of its ability to operate from very confined areas. The aircraft can be equipped with full radio and navigational equipment, night flying equipment and deicing system. Dual control can be fitted if required.

Passenger, agricultural and ambulance versions are in production and by early 1967 at least 200 had been built. Tabso has the An-14 LZ-TED and others are believed to have been exported to the German Democratic

SSSR-81550, production An-14, with flaps partly lowered.

Republic, Mongolia and Yugoslavia. At the Domodedovo display in July 1967 a military An-14 was on view. This was painted brown overall, bore the military red star on its fins and the number 06 on the fuselage.

When the production version of the An-14 appeared it was reported to bear the designation An-14A, but in May 1967 Oleg Antonov stated that the An-14A was to be a larger version with seating for 11–15 passengers. French Turbomeca Astazou airscrew-turbines have been chosen for this version, which was due to make its first flight in October 1967.

In the winter of 1958–59 the Chinese State Aircraft Factory at Shenyang completed its Capital No.1 Sha-Tu which was a four-passenger light-weight scaled-down version of the An-14. The Capital No.1 Sha-Tu is powered by two Chinese-built 160 hp M-11FR five-cylinder air-cooled radial engines with helmeted cowlings. It is possible that this Chinese aircraft may bear a close resemblance to the original An-14 design when it was intended to carry only three passengers.

Production An-14

Span 21·99 m (72 ft 1¾ in); length 11·36 m (37 ft 3¼ in); height 4·63 m (15 ft 2¼ in); wing area 39·72 sq m (427·54 sq ft); aspect ratio 12·15; track 3·6 m (11 ft 9¾ in); wheel-base 3·65 m (11 ft 11¾ in); ground clearance 66 cm (2 ft 1·9 in).

Empty weight 2,600 kg (5,732 lb); maximum fuel 278 kg* (612 lb); maximum payload 720 kg (1,587 lb); normal take-off weight 3,450 kg (7,606 lb); maximum take-off and landing weights 3,600 kg (7,936 lb).

Maximum cruising speed 180 km/h (97·13 kt) (111·85 mph); economic cruising speed at 2,000 m (6,561 ft) 175 km/h (94·43 kt) (108·73 mph); landing speed 80 km/h (43·16 kt) (49·71 mph).

Take-off run at maximum weight in still air on unpaved surface 100–110 m (328–360 ft); landing run on unpaved surface 110 m (360 ft).

Service ceiling 5,000 m (16,404 ft).

Maximum payload range with 30 min fuel reserve 180 km* (97 n.miles); range with 570 kg (1,256 lb) payload and 30 min fuel reserve 650 km* (350 n.miles).

* Soviet figures for fuel capacity and payload/range are not compatible for the quoted cruise fuel consumption of 130 kg/hr (286 lb/hr) at 175 km/h (94·43 kt) (108·73 mph) at 2,000 m (6,561 ft).

Speed during spraying 140 km/h (75·54 kt) (86·99 mph); width of spray band 60 m (196 ft) for dusting, 50 m (164 ft) for sprinkling and 30 m (98 ft) for fertilizing.

The figures quoted here for the production An-14 were published by Aviaexport and show inferior performance to that in an earlier An-14 brochure which quoted payload/ranges with 45 min fuel reserve as 300 km (161 n.miles) with 730 kg (1,609 lb) payload, 630 km (339 n.miles) with 600 kg (1,322 lb) and 730 km (393 n.miles) with 570 kg (1,256 lb). Maximum payload was then shown as 730 kg (1,609 lb) and maximum cruising speed as 190 km/h (102·52 kt) (118 mph).

The An-14A with 11 passengers has an estimated take-off run of 150 m (492 ft). With 15 passengers take-off run is estimated as 210 m (688 ft).

The An-22, SSSR-46191, at Le Bourget on arrival for the 1965 Paris Aero Show. *(Bristol Siddeley.)*

Antonov An-22

The Soviet Union is a country which is faced with the problem of having to move a considerable volume of cargo over long distances while at the same time suffering from a marked shortage of road and railway capacity. Thus it has made sense to develop air cargo to the greatest extent, and much of it is carried on scheduled passenger and cargo services. The An-10 series does provide useful capacity; nevertheless, there still is a requirement for an aircraft capable both of carrying very heavy and bulky loads over long distances and of operating from unpaved surfaces of restricted length.

Oleg Antonov's bureau was given the task of designing a very large cargo aeroplane to meet these requirements. Design studies were begun early in 1962 and resulted in the An-22, which made its first flight on 27 February, 1965.

Outside of the Soviet Union nothing was known of the An-22 until it arrived at Le Bourget on 15 June, 1965, for exhibition in the Paris Aero Show. The example shown bore the registration SSSR-46191 and was almost certainly the first prototype. It bore Aeroflot markings and on its nose the name *Antei* (*Antæus*, the giant son of Poseidon), and for a while the An-22 was known, at least in the West, by that name.

The An-22 SSSR-46191 at Le Bourget. Aeroflot's name and crest appear on the front fuselage as well as the name *Antei*.

When the An-22 appeared at Paris it was widely acclaimed by western newspapers both as a military prototype and as a 720-seat passenger aircraft. There was in fact no intention of operating the original aircraft in the passenger role, but it was stated in Paris that design work was in hand for a double-deck 724-passenger version which would be 15 m (49 ft 2½ in) longer and that it would appear in about two years' time—i.e. 1967.

The An-22 certainly seemed to have military potential but a study of Soviet cargo transport needs tends to confirm that the aircraft was designed primarily for Aeroflot cargo operations.

After the Paris Aero Show nothing more was heard of the An-22 until 27 October, 1966, when Ivan Davydov and a crew of six took the An-22 SSSR-56391 to an altitude of 6,600 m (21,653 ft) while carrying an 88,103 kg (194,234 lb) payload. At the 1967 Paris Aero Show, in May and June, this An-22 was exhibited and it was said to be the third of five prototypes then in existence. Antonov said that the first pre-production aircraft was already flying and that two An-22s were undergoing trials with Aeroflot. He also said that Aeroflot had no requirement for the 724-seat version and that the project had been shelved, but that a new version is being developed for Aeroflot's European routes. This is to be capable of carrying 300–350 passengers and 30,000 kg (66,139 lb) of cargo over a 3,000 km (1,618 n.mile) stage.

This view of the An-22 SSSR-46191 shows clearly the marked anhedral on the outer wings and also its enormous size.

The An-22 SSSR-46191 at Le Bourget. Clearly seen in this photograph are the undercarriage housings, each with a forward intake, the navigator's cabin and the contra-rotating airscrews. (*M. Stroud.*)

This view of the An-22 SSSR-46191, at Sheremetyevo in 1965, shows the twin fins and the flat underside with large loading doors. In the background are four Tu-104s and an Ilyushin Il-14.

On 9 July, 1967, a large display of Soviet civil and military aircraft was staged at Moscow's Domodedovo Airport, and four An-22s took part. SSSR-76591 was in the static exhibition, and three, with military markings, took part in the flying display during which they landed and unloaded tracked carriers on which were mounted battlefield missiles.

According to Oleg Antonov the An-22 should enter service with Aeroflot in about mid-1968 and initial annual production is likely to be 30 units.

The An-22 is a high-wing cantilever monoplane and in appearance is like a very much enlarged An-12. Unlike the earlier Antonov heavy transports, the An-22 has twin fins and rudders, this layout being chosen to overcome flexing of the rear fuselage with a single fin. The flexing problem was caused by the large under-fuselage doors, but it is not clear whether the problem was discovered during early flight trials or before construction of the prototypes.

Metal bonding and welding are extensively used in production of the An-22, and 75,000-ton presses are used to produce panels of up to 15 m (49 ft 2½ in) in length. The wing is built in seven sections, tapers in chord

Main undercarriage, undercarriage housing and starboard 15,000 hp Kuznetsov NK-12MV engines of the An-22 SSSR-46191. The radar housing can be seen beneath the entrance door. (*M. Stroud.*)

and thickness and has anhedral throughout the span although most marked on the outer third. Double-slotted flaps, in two sections each side, extend from wing roots to the ailerons which are themselves in three sections.

The fully-pressurized fuselage is a semi-monocoque structure of circular section to a point slightly aft of the wing. The rear fuselage is upswept to the tail and its underside is flattened to incorporate a large two-piece door. Three heavy frames pick up the wing spars and also form the attachment points for the levered-suspension multi-wheel main undercarriage. Nose-wheel stresses are taken through a fourth heavy fuselage frame.

The cargo area of the fuselage measures 33 m (108 ft 3 in) in length and

Tail unit of the An-22 SSSR-46191. (*John Stroud.*)

4·4 m (14 ft 5¼ in) in width and height. The floor is a reinforced titanium structure with non-slip surface and incorporates mechanical cargo locks. The very large door is in two pieces and is hydraulically operated. The aft section of the door folds upward and has on its outside surface two sets of rails which, in the raised position, form an extension of the two roof rails of the electric winch loading system which has a total capacity of 10 tons. The forward section of the door can be lowered to any desired angle so that it forms a ramp or a loading platform to meet any truck height. The doors can be opened in flight for parachute dropping of cargo.

The front bulkhead contains two doors giving access to a cabin for 28–29 cargo attendants and to the flight deck which is on a higher level. Normal crew comprises two pilots, engineer, navigator and radio operator.

The An-22 SSSR-56391 at the 1967 Paris Aero Show. This is believed to be the third prototype. (*John Stroud.*)

The An-22 SSSR-46191. (*J. M. G. Gradidge.*)

There are three circular windows in each side of the fuselage aft of the wing, and in the front fuselage there are five windows each side. There is a navigator's cabin in the extreme nose.

The tapered tailplane and divided elevators are all-metal structures carried on top of the fuselage and there are twin fins and rudders. Most of the fin area is forward of the tailplane, the fins are surmounted by slender bullet fairings and all control surfaces have trim tabs.

The An-22 has been designed to operate from a wide range of surfaces including water-sodden grass. The main undercarriage consists of three twin-wheel units in tandem on each side which retract into very large fairings on the sides of the fuselage. Each fairing contains an entrance door and steps leading to the main hold. The main wheels retract backwards and the twin nosewheels are forward retracting. For operations from surfaces of varying bearing strength, tyre pressure can be controlled in flight. There is a large air intake in the nose of each undercarriage fairing,

The An-22 SSSR-56391 on a test flight.

and the starboard fairing is known to contain an auxiliary power unit at its forward end; it is possible that the port fairing also contains an APU. The weather-warning radar is carried below the starboard undercarriage fairing.

The engines in the prototype An-22s are four 15,000 ehp Kuznetsov NK-12MV turbines with single 14-stage axial-flow compressor and five-stage turbine. This engine measures 6 m (19 ft 8 in) in length, has a diameter of 1·15 m (3 ft $9\frac{1}{4}$ in) and a dry weight of 2,300 kg (5,070 lb). Each engine drives a pair of 5·6 m (18 ft $4\frac{1}{2}$ in) diameter four-blade contra-rotating airscrews fitted with large spinners. Production aircraft will have 15,000 ehp Kuznetsov NK-12MA engines and 6·2 m (20 ft 4 in) diameter airscrews.

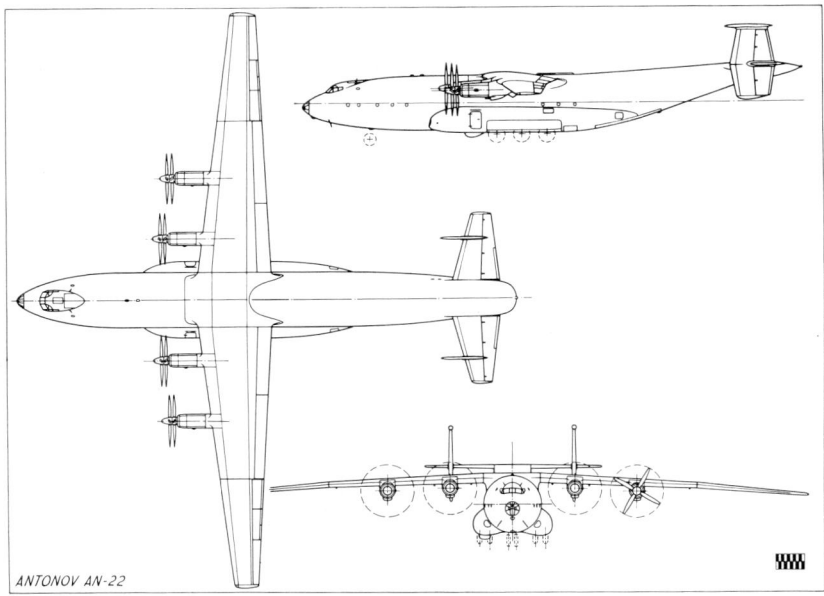

ANTONOV AN-22

The An-22 can carry standard containers for transhipment to rail, sea or river transport, while its large capacity and unobstructed loading area makes possible the carriage of drilling, earth-moving and building equipment, bridge units, generators, tractors, railway wagons and buses. It is thought that Aeroflot has already used An-22s for experimental carriage of fish and fruit and undertaken trials in a wide range of climates.

When it appeared the An-22 was, in terms of maximum weight, the biggest aircraft ever produced.

The abandoned 724-seat An-22 variant would have had accommodation for 423 passengers in six cabins on the upper deck and 301 passengers in four cabins on the lower deck. There would have been bar and buffet areas on each deck and a total of 120 cu m (4,237 cu ft) baggage capacity. One cabin would have been used for mothers and children, and there would have been provision for television and the showing of films. Each deck would have had three entrance doors on each side and the centre doors of the lower deck would have been in the undercarriage fairings. Six stairways would have connected upper and lower decks.

Span 64·4 m (211 ft 3½ in); length 57·31 m (188 ft 0¼ in); height 12·53 m (41 ft 1¼ in); wing area 480 sq m (5,166·68 sq ft). Empty equipped weight 114,000 kg (251,327 lb); fuel 43,000 kg (94,799 lb); maximum payload 80,000 kg (176,370 lb); maximum take-off weight 250,000 kg (551,160 lb). Maximum speed 740 km/h (399·3 kt) (459·81 mph); cruising speed 560–640 km/h (302·18–345·34 kt) (347·97–397·68 mph); take-off run 1,100–1,300 m (3,608–4,265 ft); landing run 800 m (2,624 ft); maximum payload range 5,000 km (2,698 n.miles); maximum fuel range 11,000 km (5,936 n.miles) with 45,000 kg (99,208 lb) payload.

Antonov An-24

Although Aeroflot had introduced its turbojet Tu-104s in 1956 and its large airscrew-turbine powered An-10s and Il-18s in 1959, as late as 1962 more than 140 Federal routes were being operated by Il-14s, another 10 by Il-12s, more than 60 by Li-2s, and these rather elderly piston-engined aircraft were also still in largescale use on local services. Nevertheless, steps had been taken to develop a replacement aircraft and re-equipment began in 1962.

The first prototype An-24, SSSR-L1959. Later aircraft had a deepened dorsal fin, lengthened engine nacelles and a ventral fin.

The first prototype An-24 with undercarriage and flaps lowered.

Aeroflot had a requirement for a 32–40 seat aircraft for short and medium stage routes and this type would have to be capable of operation in a wide range of climates and work from small unpaved aerodromes. The new aircraft had to be reliable, safe and economic, as well as providing modern standards of comfort on the regional and local services.

The task of designing this new type was entrusted to the design bureau at Kiev under the leadership of Oleg K. Antonov and work on the project began in December 1957. During the development stage the seating requirement was increased to 44 and, later, to 48.

The aircraft produced by Antonov was the An-24, an all-metal high-wing cantilever monoplane powered by two airscrew-turbines and generally resembling in appearance, size and weight the Handley Page Herald and Fokker F.27 Friendship.

The first prototype, appropriately registered SSSR-L1959, made its first flight on 20 December, 1959, when it was flown by Yuri Kurlin and G. Lysenko. It appears that some stability problems were encountered because when the second prototype appeared, SSSR-L1960, it incorporated a number of modifications. These included deepening of the dorsal fin, slight changes in the rear fuselage, addition of a ventral fin, aft extension of the engine nacelles and slight lengthening of the nose. In the first prototype the flaps extended from ailerons to fuselage and the rear of the nacelles drooped with the flaps, but in the second prototype and in production aircraft the flaps were divided by the nacelles.

The prototypes were followed by five pre-production aircraft, three for flight test and two for static and fatigue testing. These tests resulted in the An-24 being cleared for a life of 30,000 hr.

After more than two years of flight tests deliveries began, with the first examples going to the Ukrainian Directorate of Aeroflot in April 1962. Crew training began immediately and, after deliveries to the Moscow

The An-24V SSSR-46791 taking part in the 1965 Paris Aero Show. The markings were in two shades of blue and the top of the nose was grey. (*Butler-Green Aviation Photo Service.*)

Directorate, crews were trained for the White Russian, Ural and Northern Caucasia Directorates.

Cargo services over the Kiev–Nikolayev–Kherson route began in July 1962, a technical proving flight with passengers was made between Kiev and Krasnodar that September, and at 13.15 hr Moscow time on 31 October, 1962, an An-24 left Kiev for Kherson on the first scheduled passenger service to be operated by the type. The An-24 was rapidly put into service on other routes linking the cities of the Ukraine. In the first year of operation the Ukrainian An-24s averaged 1,455 hr utilization, and their largescale use in the Ukraine is illustrated by the fact that by mid-1967 they had carried a total of $2\frac{1}{2}$ mn passengers plus 100,000 tons of cargo.

On 1 December, 1963, An-24s began working the Moscow–Saratov route, and it is known that in 1964 the type went into operation on the following routes: Moscow–Cheboksary on 5 February; Moscow–Gudauta, Moscow–Kursk and Moscow–Tula on 10 June; Moscow–Makhachkala on 24 August; Krasnoyarsk–Kyzyl on 6 September; and Moscow–Voronezh–Volgograd–Elista–Groznyy on 28 October. During that summer it had also been introduced on the Moscow–Kazan'–Perm route.

The first production batch of An-24s were of the basic model with accommodation for 44 passengers, but production was soon switched to the An-24V with higher operating weights and accommodation for up to 50 passengers. The An-24V, SSSR-46791, was shown at the 1965 Paris Aero Show and caused some confusion by bearing on its nose the designation An-24B in cyrillic characters. It seems likely that most of the An-24s introduced in 1964 were of this later model.

Apart from proving a most useful aircraft for Aeroflot, the An-24 attracted export orders. By the end of 1964 Lebanese Air Transport (later Lebanese Overseas Airways) had taken delivery of OD-AEN. Misrair took delivery of the first of its original order for seven An-24Vs on 5 August, 1965, and on the following day this aircraft, SU-ANV, began operating the Cairo–Luxor–Aswan service, crew training already having taken place in the USSR. Polskie Linie Lotnicze (LOT) took delivery of the first three of its order for 10 in March and April 1966 and on 20 April that year began An-24 services when SP-LTA operated Warsaw–Wrocław and Wrocław–Warsaw services. Other orders were: Air Guinée three, Air Mali two, Cubana eight, Interflug six, Mongolian Airlines three, Pan-African Air Services (Tanzania) two, Tabso seven and Tarom three. Syrian Arab Airlines was reported to have ordered three, but did not take delivery. The Polish Air Force also uses the An-24.

By the summer of 1965 An-24s were in service with at least nine Aeroflot Directorates and Aviation Groups and were working more than 50 Federal routes. A year later An-24s and An-24Vs were in operation with at least 10 Directorates and with 235 Division. They were then reported to be working on more than 150 routes of which one-third were Federal, and during 1966 carried about 3,300,000 passengers. Some cargo services were also being flown by An-24s.

Although standard accommodation is for 44 passengers in the An-24 and 50 in the An-24V, four seating configurations were known to be in use with Aeroflot in 1967. The Kazakh, Moscow, Volga and Western Siberia Directorates used the 44-seat layout and the Azerbaydzhan, Georgian, Latvian, North, Turkmen and White Russian Directorates had 50-seat aircraft. Krasnoyarsk, Northern Caucasia, Ukrainian, Ural and Uzbek Directorates employed a 48-seat layout, the Ukrainian Directorate having increased the seating from 44 to 48. The fourth configuration was a 36-seat layout used by 235 Division on flights from Moscow to Khar'kov and Gelendzhik. Polskie Linie Lotnicze and Mongolian Airlines both employed 44-seat An-24s.

Production was at the rate of four a month by mid-1966 and it appears that about 140 An-24s had been produced by the end of that year.

In the first half of 1967 it was announced that, beginning in January 1968, the An-24 Series II would be produced, and the earlier aircraft were retroactively designated Series I. The Series II An-24s have increased

Polskie Linie Lotnicze (LOT) An-24V SP-LTG. Markings are dark blue with the red and white national colours on the nose and fin and rudder. In the background is a LOT Il-18. (*Polskie Linie Lotnicze.*)

Lebanese Air Transport's An-24 OD-AEN. (*Courtesy William Green.*)

Interflug's An-24V DM-SBG at Leipzig. Fuselage and tail markings are red. (*Interflug*.)

The Misrair An-24V SU-AOC at Cairo Airport. This aircraft has twin ventral fins and is believed to have uprated engines for operation in high temperatures. Fuselage, engine nacelle and tail markings are red, white and black. (*Misrair*.)

power and there are three main variants, the An-24V Series II, the An-24TV cargo aircraft and the An-24RV with an RU-19-300 auxiliary turbojet in the starboard nacelle. Prototypes of the An-24TV, SSSR-46280, and An-24RV, SSSR-98104, were exhibited at the Paris Aero Show in May and June 1967.

The wing of the An-24 is a two-spar structure and comprises centre section with two outer sections on each side. The parallel-chord centre section is untapered in thickness, without dihedral, and at its extremities carries the engine mountings. Outboard of the engines the wing is tapered in plan and thickness and at quarter chord the leading-edge sweepback is $6\frac{1}{2}$ deg. The outermost sections have 2 deg anhedral. The skin is attached to the spars and ribs by electro spot-welding. Hydraulically-operated double-slotted Fowler-type camber-changing and area-increasing flaps extend from the engine nacelles to the ailerons and between the nacelles and the fuselage. Flap chord is 36 per cent of the inboard wing chord. The ailerons are each in two sections and carry fibre-glass trim tabs.

The fuselage is a semi-monocoque structure of circular section, with increased radius on the underside. Electro spot-welding and metal-to-metal bonding are used to attach the stringers to the fuselage skin. Except for the extreme nose and the tail cone, the fuselage is pressurized to a differential of 0·3 kg/sq cm (4·26 lb/sq in) at an altitude of 6,000 m (19,685 ft).

The fin and tailplane are two-spar all-metal structures, the former having 21 deg 30 min sweepback and the latter slight dihedral. The control surfaces are metal skinned, the elevators have cable-operated tabs and the rudder incorporates electrically-operated trim and servo tabs.

All undercarriage units have twin wheels, are hydraulically retracted forward and have free fall. Main-wheel tyre pressure is 5 kg/sq cm (71 lb/sq in) and nosewheel tyre pressure 3·5 kg/sq cm (50 lb/sq in), but for operation from sandy or wet unpaved surfaces these pressures can be lowered to 3·5 kg/sq cm (50 lb/sq in) and 2·5 kg/sq cm (35·5 lb/sq in) respectively.

The Series I aircraft are powered by two Ivchenko AI-24 airscrew-turbines with 10-stage axial-flow compressors and three-stage turbines. The AV-72 four-blade airscrews are fully-feathering and have a diameter of 3·9 m (12 ft $9\frac{1}{2}$ in). Normal rated power is 2,100 ehp and take-off power is 2,550 ehp dry or 2,800 ehp with water injection. Cruise fuel consumption at 6,000 m (19,685 ft) is 340 kg/eng/hr (750 lb/eng/hr). Normal fuel capacity is 4,060 kg (8,950 lb) carried in four centre-section flexible bag tanks and integral tanks in the inner sections of the outer wings. Fuel capacity can be increased to 4,820 kg (10,626 lb) by the fitting of an additional four centre-section bag tanks. Each group of tanks can be gravity filled through the upper surface of the wing but there is a pressure refuelling point on the inner side of the starboard nacelle. A TG-16 turbo-generator for ground power can be installed in the rear of the starboard nacelle at customer's request.

The Series II aircraft have Ivchenko AI-24T engines each developing 2,820 ehp, and the An-24RV has a 900 kg (1,984 lb) thrust Tumanskii RU-19-300 auxiliary turbojet in the starboard nacelle in place of the optional TG-16 APU. The RU-19-300 improves performance and controllability in flight in case of an engine failure, and its use on take-off allows maximum load to be carried from aerodromes up to an elevation of 2,500 m (8,202 ft) and in temperatures of up to ISA plus 30 deg C. (45 deg C.).

Wing, fin and tailplane leading edges and engine intakes are protected from icing by hot air bled from the 10th stage compressor of each engine. Airscrews and two main flight deck windows are electrically heated. Warm

The cabin of a Polskie Linie Lotnicze An-24V. (*Polskie Linie Lotnicze.*)

and cold air for the cabin is tapped from the compressors of both engines and use is made of heat-radiant wall panels.

The passenger cabin in the An-24 is 9·69 m (31 ft 9½ in) long, 2·76 m (9 ft 0½ in) wide and 1·91 m (6 ft 3 in) high. The passenger door, aft on the port side, is 1·4 m (4 ft 7¼ in) high and 75 cm (2 ft 5½ in) wide, and the forward loading door, on the starboard side, measures 1·1 m by 1·2 m (3 ft 7¼ in by 3 ft 11¼ in). Sill heights are 1·4 m and 1·3 m (4 ft 7¼ in and 4 ft 3 in) respectively.

The Series I An-24V was offered in five configurations. The main passenger type had 50 seats arranged in 12 rows with double seats on each side and two aisle seats at the rear of the cabin. Between the aft seats and the cabin walls were carry-cots for infants. Forward of the cabin was a baggage compartment and a coat cupboard, and aft of the cabin was a buffet, toilet and more coat space. A mixed passenger–cargo version provided accommodation for 30 passengers aft and 14 cu m (494

An-24TV cargo aircraft SSSR-46280 at the 1967 Paris Aero Show. On the left is the Tu-134 SSSR-65610 and on the right an Il-62. (*John Stroud.*)

cu ft) of cargo space forward, and an all-cargo version had a cargo volume of 35 cu m (1,236 cu ft). A salon, or executive, version of the An-24 provides seats for 16 arranged in facing pairs divided by tables, and has a sleeping cabin aft of the main cabin. The fifth version is a de luxe layout for eight passengers and has two separate single-berth sleeping cabins. In the all-passenger aircraft seat pitch is 720 mm (28·2 in) with 50 seats, 750 mm (29·5 in) with 48 seats and 810 mm (31·8 in) with 44 seats.

The Series II An-24V is offered in 46 and 50 seat all-passenger versions, in a mixed configuration with cargo forward and 18 seats aft, and in a de luxe version with 18 seats.

The An-24TV at Paris in 1967. The twin ventral fins can be seen and, behind the figures, the loading door. The upper fuselage and fin were yellow and the fuselage stripes grey, black and white. The rudder was white. (*John Stroud.*)

The An-24TV cargo aircraft has a hold length of 11·1 m (36 ft 5 in) and this is equipped with a load conveyor system in the floor. A rear inward-opening underfloor loading door measures 2·85 m (9 ft 4¼ in) in length and has maximum and minimum widths of 1·4 m and 1·1 m (4 ft 7¼ in and 3 ft 7¼ in). A hoist capable of lifting 1,500 kg (3,306 lb) can be used for lifting loads through the rear hatch. Fold-up seats along the cabin walls can be used to carry up to 40 passengers. In place of the nine circular windows in each side of the passenger aircraft, the An-24TV has four windows in the starboard side and three in the port side. There are also two angled-out ventral fins instead of the standard single fin of other versions. These twin ventral fins are also fitted to Misrair's An-24s which are believed to have uprated engines.

An-24s are equipped with weather-warning radar. Normal crew comprises two pilots and a cabin attendant, but the Series II aircraft can be equipped with seats for two pilots, navigator on the port side, radio operator to starboard, and a removable centrally-located engineer's seat.

An-24s on the production line at Kiev.

Span 29·2 m (95 ft 9½ in); length 23·53 m (77 ft 2¼ in); height 8·32 m (27 ft 3½ in); wing area 72·46 sq m (779·95 sq ft); aspect ratio 11·7; track 7·9 m (25 ft 11 in); wheelbase 7·85 m (25 ft 9 in).

An-24V Series I

Empty weight 13,300 kg (29,321 lb); normal fuel 4,060 kg (8,950 lb); maximum fuel with additional centre-section tanks 4,820 kg (10,626 lb); maximum payload, passenger version, 5,500 kg (12,125 lb); maximum payload, cargo version, 5,700 kg (12,566 lb); normal take-off weight 20,000 kg (44,092 lb); maximum take-off weight 21,000 kg (46,297 lb); maximum landing weight 21,000 kg (46,297 lb).

Cruising speed at recommended cruise level of 6,000 m (19,685 ft) 450–500 km/h (242·82–269·8 kt) (279·62–310·69 mph); landing speed at 19,000 kg (41,888 lb) 165 km/h (89 kt) (102·51 mph).

Take-off run at 19,500 kg (42,990 lb) at sea level ISA, 500 m (1,640 ft); take-off run at 21,000 kg (46,297 lb) at sea level ISA, 620 m (2,034 ft), at 2,000 m (6,561 ft) elevation ISA, 940 m (3,083 ft); take-off distance to 15 m (49 ft) at 19,500 kg (42,990 lb) at sea level ISA, 870 m (2,854 ft); take-off distance to 15 m (49 ft) at 21,000 kg (46,297 lb) at sea level ISA, 1,020 m (3,346 ft), at 2,000 m (6,561 ft) elevation ISA, 1,450 m (4,757 ft); take-off distance to 10·5 m (35 ft) at 21,000 kg (46,297 lb) at sea level ISA with one engine inoperative, 1,730 m (5,676 ft); sea level rate of climb at 19,500 kg (42,990 lb), undercarriage and flaps retracted, 7·7 m/sec (1,515 ft/min); time to 6,000 m (19,685 ft) from unstick, at 19,500 kg (42,990 lb), 17·5 min.

Landing run at 19,000 kg (41,888 lb) at sea level ISA, 590 m (1,935 ft); landing distance from 15 m (49 ft) at 19,000 kg (41,888 lb) at sea level ISA, 1,130 m (3,707 ft).

Service ceiling at 21,000 kg (46,297 lb) 9,000 m (29,527 ft); guaranteed single-engine ceiling at maximum weight 2,000 m (6,561 ft).

Maximum payload range, passenger version, at 6,000 m (19,685 ft) and 475 km/h (256·31 kt) (295·15 mph), with 45 min fuel reserve, 650 km (350 n.miles); maximum payload range, cargo version, at 8,000 m (26,246 ft) and 475 km/h (256·31 kt) (295·15 mph), with 45 min fuel reserve, 650 km (350 n.miles); maximum fuel range, with additional tanks and 27 passengers, at 6,000 m (19,685 ft) and 475 km/h (256·31 kt) (295·15 mph), 2,500 km (1,349 n.miles); maximum fuel range, with additional tanks and 2,700 kg (5,952 lb) cargo, at 8,000 m (26,246 ft) and 475 km/h (256·31 kt) (295·15 mph), 2,900 km (1,564 n.miles).

Take-off and landing performance quoted is for dry concrete runway. Landing run and distance is with wheel and airscrew braking.

With 1,000 kg (2,204 lb) payload and one hour's fuel the An-24V can take-off with a ground run of only 150 m (492 ft).

An-24V Series II

Maximum fuel with additional centre-section tanks 4,800 kg (10,582 lb); maximum payload 5,700 kg (12,566 lb); maximum take-off weight at sea level ISA 21,000 kg (46,297 lb), at sea level ISA plus 30 deg C. and without water injection 18,800 kg (41,446 lb), at sea level ISA plus 30 deg C. with water injection 20,700 kg (45,635 lb).

Cruising speed 450–500 km/h (242·82–269·8 kt) (279·62–310·69 mph).

Take-off run 600 m (1,968 ft); single-engine sea level rate of climb at maximum weight ISA, 1·9 m/sec (374 ft/min); single-engine sea level rate of climb at maximum weight ISA plus 30 deg C. and without water injection, 1·5 m/sec (295 ft/min); single-engine sea level rate of climb at maximum weight ISA plus 30 deg C. with water injection, 5·2 m/sec (1,023 ft/min).

Landing run 550 m (1,804 ft).

Service ceiling 8,400 m (27,559 ft).
Range with 5,500 kg (12,125 lb) payload and 580 kg (1,278 lb) reserve fuel in ISA, 550 km (296 n.miles); range with 3,300 kg (7,275 lb) payload and 580 kg (1,278 lb) reserve fuel in ISA plus 30 deg C., 550 km (296 n.miles); maximum range with 45 min fuel reserve 2,400 km (1,295 n.miles).

An-24RV

Maximum take-off weight at sea level up to ISA plus 30 deg C., 21,800 kg (48,060 lb); single-engine sea level rate of climb at maximum weight ISA, 3·4 m/sec (669 ft/min); single-engine sea level rate of climb at maximum weight ISA plus 30 deg C., 2·4 m/sec (472 ft/min); range with 5,500 kg (12,125 lb) payload and 580 kg (1,278 lb) reserve fue up to ISA plus 30 deg C., 550 km (296 n.miles).

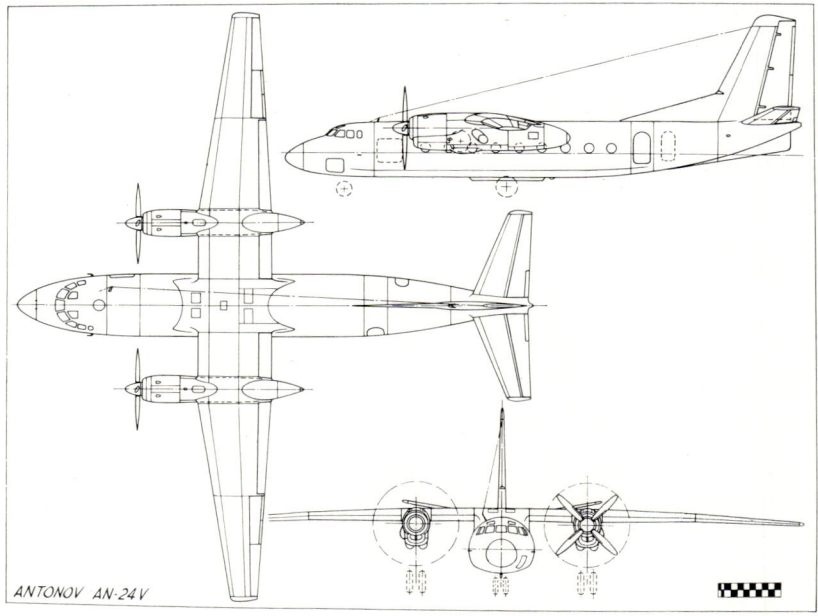

Known dates for An-24 introductions and technical proving flights:

Aeroflot

1962 July	Kiev–Nikolayev–Kherson (cargo services)
1962 September	Kiev–Krasnodar (technical flight, with passengers)
1962 October 31	Kiev–Kherson (first An-24 passenger services)
1963 December 1	Moscow–Saratov
1964 February 5	Moscow–Cheboksary
1964 June 10	Moscow–Gudauta, Moscow–Kursk and Moscow–Tula
1964 August 24	Moscow–Makhachkala
1964 September 6	Krasnoyarsk–Kyzyl
1964 October 28	Moscow–Voronezh–Volgograd–Elista–Groznyy
1964	Moscow–Kazan'–Perm

1965 January 3	Moscow–Izhevsk–Tyumen
1965 January 26	Moscow–Kazan'–Tyumen
1965 March 15	Moscow–Nikolayev
1965 October 1	Ashkhabad–Mary
1965	Riga–Vil'nyus–L'vov–Kishinev
1967	Yerevan–Tbilisi

Air Mali

1968 April 15	Bamako–Abidjan (by TZ-ACT)

Interflug

1966 March 20	Berlin–Dresden (believed first Interflug An-24 service)

Misrair

1965 August 6	Cairo–Luxor–Aswan (first Misrair An-24 service, by SU-ANV)

Mongolian Airlines

1965–66 winter	Ulan Bator–Irkutsk

Polskie Linie Lotnicze (LOT)

1966 April 20	Warsaw–Wrocław (first LOT An-24 service, by SP-LTA)

Tabso

1967 January	Sofia–Istanbul (unconfirmed)

Beriev Be-6 and Be-10

Details of civil operation of marine aircraft in the Soviet Union are almost completely unknown. Schavrov's Sch-2 single-engine two-seat amphibian is known to have been used before the war on passenger and mail services to remote areas which could not even be served by the U-2 landplanes. The much later Sch-7, of which there was only one, was used to serve towns along the Volga between Saratov and Astrakhan. MP-1 single-engine flying-boats were used by Aeroflot, and Dornier Wals and MP-7s (Soviet transport version of the Consolidated Catalina) were also used on Soviet transport operations. It is even reported that Aeroflot used the Douglas DF-151, Martin 156 and Sikorsky S-43A, the Martin being given the designation PS-30.

Since the war An-2 biplanes on floats have been put into service, but nothing whatever has been released by the USSR about civil operations by flying-boats; yet there is some evidence that two types have been used. Whether operation of flying-boats is the responsibility of Aeroflot or of Glavsevmorput (Administration of Northern Sea Routes) is not known, neither is the extent of Northern Sea Routes air services known. The 1963 edition of the *Oxford Regional Economic Atlas of the USSR and Eastern Europe* shows a number of routes said to be operated by Glavsevmorput, and several of these follow some of the main rivers including the Ob,

Yenisey and Lena. Others appear to follow lesser rivers or serve communities close to lakes. North coast ports are served, as well as Arctic stations on Novaya Zemlya and Severnaya Zemlya, but because of ice it seems likely that the most northern points are served by landplanes equipped with skis. The southern limits of these routes terminate at points served by Aeroflot's normal Federal services.

In 1957 a number of Beriev Be-6 flying-boats with civil markings were seen on Khimki lake, north of Moscow, and on 31 December, 1965, a set of Soviet civil aviation stamps was issued with one stamp depicting a Beriev Be-10 in Aeroflot markings. This stamp is extremely interesting because the Be-10 is turbojet-powered and if it is really in airline service it is the only jet flying-boat ever to be used for civil airline operations.

Both the Be-6 and Be-10 were designed by the bureau headed by Georgii Mikhailovich Beriev at Taganrog on the Sea of Azov. After the war the Morskaya Aviatsiya (Soviet Marine Aviation) required a modern marine reconnaissance flying-boat with long range, and to meet this specification Beriev designed the LL-143 (Letayushchaya Lodka—flying-boat). This was a large twin-engine aircraft with gull wing, single-step hull, non-retractable wing tip floats and twin fins and rudders. The wing had a straight trailing edge but marked taper on the leading edge. There was considerable dihedral between the hull and the engines and slight anhedral outboard. Full-span flaps and ailerons were fitted.

The hull of the LL-143 was quite deep, housed a normal crew of eight and was roomy enough to accommodate a relief crew. Armament comprised a 23-mm cannon in the bow and a pair of these guns in tail and dorsal turrets. The tail turret was a remote-controlled barbette. Mines, depth-charges or torpedoes were carried on underwing pylons outboard of the engines.

The dihedral tailplane had sweepback on its leading edge and the fins and rudders were approximately oval.

The two closely-cowled Shvetsov ASh-72 eighteen-cylinder air-cooled radial engines each developed 2,000 hp and drove four-blade airscrews.

A Soviet Marine Aviation Be-6. Some were seen, in 1957, with civil markings but no photograph has been traced. (*Courtesy William Green.*)

A Be-10 turbojet flying-boat of Soviet Marine Aviation. A Be-10 with Aeroflot markings appeared on a Soviet civil aviation postage stamp in 1965. (*Courtesy William Green.*)

The LL-143 passed its manufacturer's and State trials in 1947 and was ordered into production as the Be-6. The Be-6 was flown, by M. I. Tsepilov, for the first time in 1949 and it incorporated a number of modifications although still generally resembling the LL-143. The engines were 2,300 hp ASh-73TK radials, the forward part of the hull was redesigned and the bow gun omitted. A magnetic anomaly stinger replaced the tail gun barbette, and a retractable radome was built into the hull aft of the step. The Be-6 went into widescale service as a maritime reconnaissance and patrol aircraft, and it is known to have been used for transport duties although details of the accommodation are not known.

In 1949 the Soviet Union's first turbojet flying-boat began its flight trials. This was the Beriev Be-R-1 gull-wing monoplane powered by two 2,740 kg (6,040 lb) thrust Klimov VK-1 centrifugal-flow turbojets. The hull had a length-to-beam ratio of about 1:8, there were outward retracting wing tip floats and the single fin and rudder carried a high-mounted tailplane. Remote-controlled tail armament was fitted. Flight testing continued until 1951. The Be-R-1 achieved a speed of 770 km/h (415·49 kt) (478·46 mph) but was not put into production as there was no operational requirement for this category of aircraft. The dimensions and weights of the Be-R-1 are not known.

The Be-R-1, although not proceeded with, did provide valuable experience which was used in the design of the Be-10, originally known as the M-10.

The Be-10 was first seen, although not identified, when it flew past at the Aviation Day display in Moscow in 1961. In August and September that year a flying-boat designated M-10 and powered by two 6,500 kg (14,330 lb) thrust AL-7 turbojets set a number of international records including a speed of 912 km/h (492·11 kt) (566·68 mph) and an altitude of 14,962 m (49,088 ft). Details of these records appear in Appendix V.

The Be-10 has a highly sweptback wing with flat centre section and prominent anhedral outboard of the engines which are beneath the wing

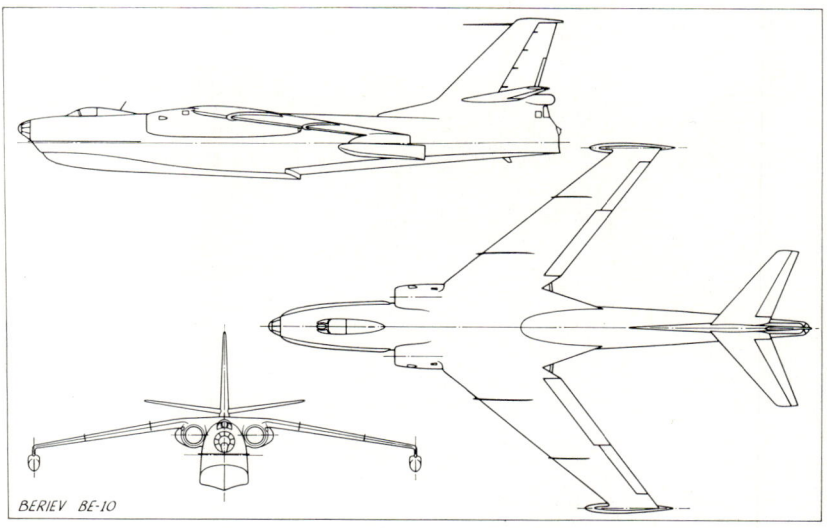

BERIEV BE-10

and close alongside the hull. The wing has two boundary layer fences on each side, wide-span flaps and at its tips non-retractable floats carried on short pylons. The single-step hull has a high length-to-beam ratio, houses the pilots in a raised canopy, and in its deep stern has a gunner's position and radar-directed 23-mm cannons.

The fin and rudder is sweptback, there is a small dorsal fin and the sweptback dihedral tailplane and elevators are mounted on the deepened rear hull just below the fin.

The turbojets are reported as the AL-7RV type and are almost certainly of Lyulka design. Spray fences on the bows keep water from entering the engine intakes.

The Be-10 is in service with Morskaya Aviatsiya.

Nothing at all is known about the transport version and the sole evidence for it is the Soviet postage stamp. As it appears to be in service with or on trial for Aeroflot it is strange that such strict secrecy has been maintained when the USSR could claim the distinction of being the only operator in the world of a jet transport flying-boat.

Be-6

Span 33 m (108 ft 3¼ in); length 23·2 m (76 ft 1¼ in). Loaded weight 23,400 kg (51,587 lb). Maximum speed 415 km/h (223·93 kt) (257·87 mph) at 2,400 m (7,874 ft); minimum speed 155 km/h (83·63 kt) (96·3 mph); range 4,800 km (2,590 n.miles).

Be-10

Dimensions unknown. Sweepback at 25 per cent chord about 48 deg. Maximum payload 15,206 kg (33,523 lb). Maximum speed 912 km/h (492·11 kt) (566·68 mph); absolute ceiling 14,962 m (49,088 ft).

No other details are known of the Be-6 and Be-10.

Beriev Be-30

It is expected that by 1970 Aeroflot will be carrying about 75 mn passengers a year and that of this total 35 mn will be flying on local services. For some years the short-stage local services, with passenger loads varying between tens and hundreds a day, have mostly been flown by the Antonov An-2 biplanes.

During 1965 Beriev's bureau was given the job of designing a safe and reliable modern aircraft to replace the An-2s on the longer of these local short-stage services. The result was Beriev's first landplane, the Be-30, which is reported to have made its first flight on 3 March, 1967. The first prototype is believed to have been SSSR-23166 and it would seem to have been completed late in 1966. The type was shown in public for the first time when another prototype, SSSR-30167, was put on view in the static aircraft park at the Domodedovo show in July 1967.

The engines of the Be-30 have only been described by Beriev as the TVD-10 free-shaft turbines of 970 ehp, but TVD are the initial letters of the Russian words for turbine, propeller and engine. It has been suggested that the Be-30 made its first flight powered by two 740 hp ASh-21 air-cooled radial piston engines but this seems unlikely. The Soviet Union has ordered a batch of French Turbomeca Astazou XII propeller-turbines for trial in the Antonov An-14 Pchelka and it is possible that a version of the Astazou may be tried in the Be-30; but the Astazou does not develop 970 hp and a Soviet drawing of the engine in the Be-30 does not resemble the Astazou.

SSSR-23166, believed to be the first prototype Be-30. Another example, seen in 1967, had modified engine cowlings and three-blade airscrews.

SSSR-23166 and SSSR-30167 are the only Be-30s which have been seen, or illustrated outside the USSR. The two aircraft had different-shaped engine cowlings, SSSR-23166 had four-blade airscrews with broad blades, and SSSR-30167 had three-blade airscrews. The braking airscrews are reported to have a diameter of 3 m (9 ft $10\frac{1}{4}$ in).

The Be-30 has been designed mainly for operation over stages of 200–800 km (108–431 n.miles) and has normal accommodation for 14–15 passengers. It has short take-off and landing ability, and the engine intakes have been kept as high as possible to avoid ingestion of dirt and stones when operating from unprepared surfaces.

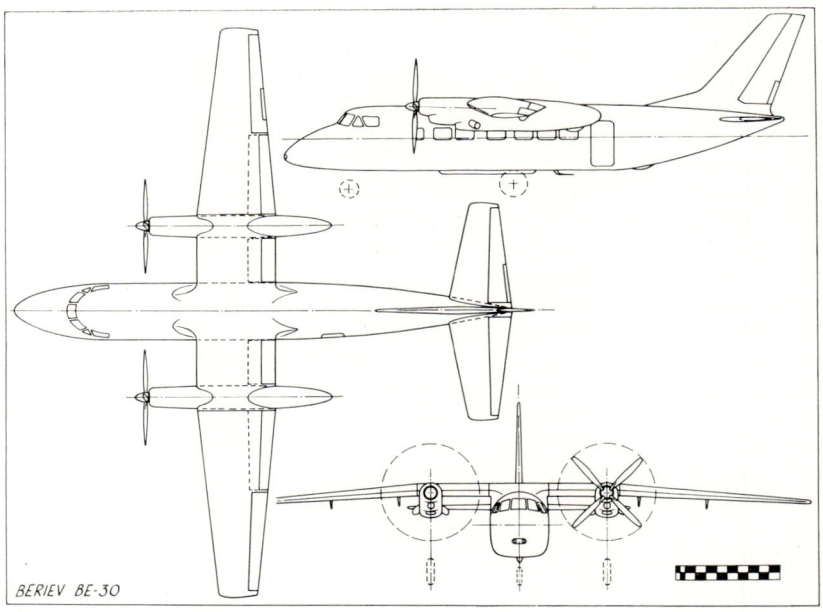

BERIEV BE-30

In layout the Be-30 is a high-wing cantilever monoplane with single fin and rudder and retractable nosewheel undercarriage. The wing is built in three sections, an uninterrupted centre section and two outer sections. The centre section is of parallel chord, untapered in thickness, and carries the engine mountings near its extremities. The outer wings have anhedral and taper in chord and thickness. Double-slotted flaps extend between engine nacelles and ailerons and between engine nacelles and fuselage. A tab is incorporated in the starboard aileron.

The fuselage is of approximately rectangular section with curved top, and the wing spars are carried above the fuselage structure to give an unobstructed cabin. The fuselage is not pressurized.

The tailplane and elevators are tapered in plan and thickness but without dihedral. The fin and rudder is sweptback and there is a small dorsal fin. The rudder and starboard elevator incorporate trim tabs, but the tailplane does not appear to have variable incidence.

All undercarriage units have single wheels with low-pressure tyres and the main units retract aft to be housed in large fairings which extend aft of the wing trailing edge.

The Be-30 has all highly-stressed skin panels produced by mechanical or chemical milling and much use has been made of metal-to-metal bonding. Honeycomb structure has been used extensively in wing and tail structure. Savings in weight and production man-hours have been

The Be-30 SSSR-30167 at the Domodedovo display in July 1967. Beyond the Be-30 is the Yak-40 SSSR-19661 and on the right is the An-2P SSSR-25584.

achieved by largescale use of fibre-glass for non-load-carrying components. Hot air tapped from the engine compressors is used for wing and tail leading edge ice protection, the airscrews have electric heating and the engine intakes are protected by hot oil.

The cabin has seven single seats on each side, there are baggage holds fore and aft and a rear lavatory. There are one small and six large windows each side. The entrance is aft of the wing on the port side and folding steps are fitted. The flight deck is equipped for one-crew operation and the right-hand seat can be used as a 15th passenger seat.

The Be-30 has a duplicated electrical system, and is said to have the necessary equipment for automatic approaches down to 30 m (98 ft). Landing and taxi-ing lights are installed in the nose. As with most Soviet transport aircraft the engines can be started with the aircraft's own power—in this case batteries.

The Russians have referred to the Be-30 as the Air Microbus. The aircraft can be quickly converted for cargo or ambulance work, and survey and executive versions may be produced.

Cost per tonne-km in kopecks is quoted as 24 over a 200 km (108 n.mile) stage with 1,300 kg (2,866 lb) payload, and 20·8 over an 800 km (431 n.mile) stage with 1,130 kg (2,491 lb) payload. Most economic operation will be over 600 km (323 n.mile) stages on which a 1,285 kg (2,832 lb) payload can be carried for 19·4 kopecks per tonne-km, the figure of 19·4 kopecks/tonne-km being approximately equal to 2/8d per ton-mile. Beriev claims that the Be-30 can operate profitably with 50 per cent payload.

Span 17 m (55 ft 9¼ in); length (fuselage) 15 m (49 ft 2½ in); height (undercarriage retracted) 4·82 m (15 ft 10 in); wing area 32 sq m (344·44 sq ft); track 5·2 m (17 ft 0¾ in); wheelbase 4·75 m (15 ft 7 in). Payload 1,300 kg (2,866 lb); maximum take-off weight 5,700 kg (12,566 lb). Cruising speed at 2,000 m (6,561 ft) 460–480 km/h (248·22–259 kt) (285·83–298·26 mph); take-off speed 135 km/h (72·84 kt) (83·88 mph); landing speed 130 km/h (70·15 kt) (80·78 mph). Take-off run at maximum weight 170 m (557 ft); take-off distance to 15 m (49 ft) about 260 m (853 ft); take-off distance to 10·6 m (35 ft) at maximum weight in ISA with one engine inoperative, 475 m (1,558 ft); take-off distance to 15 m (49 ft) at maximum weight in ISA plus 15 deg C., with one engine inoperative, 525 m (1,722 ft); balanced take-off field length 550 m (1,804 ft); landing run 130 m (426 ft). Maximum payload range 400 km (215 n.miles); range with 1,285 kg (2,832 lb) payload 600 km (323 n.miles); range with 1,130 kg (2,491 lb) payload 800 km (431 n.miles); maximum range, with 680 kg (1,499 lb) payload, 1,300 km (701 n.miles).

Ilyushin Il-12 and Il-14

Before the war the main units of Aeroflot's fleet were the ANT-9s, also known as PS-9s, some ANT-35s (PS-35s) and a few Douglas DC-3s. The DC-3 proved to be as suitable for operation in the USSR as it was elsewhere, and licence production was begun in the Soviet Union in 1939 with the type going into passenger service in 1940. In Aeroflot service the DC-3 was known as the PS-84, and this designation was also applied to the US-built C-47s which were supplied to the USSR during the war and used by the Soviet Air Force and Aeroflot. The engineer chosen to go to the United States to study the DC-3, and then make certain modifications and get it into production, was Boris Lisunov, and on 17 September, 1942, the PS-84 was redesignated Lisunov Li-2.

About 2,000 Li-2s were built and after the war these, and the modified C-47s which had their double doors removed and a single door installed on the starboard side, formed the backbone of Aeroflot's fleet. Many are still in service, with at least 22 Directorates employing them in the summer of 1967; a year earlier they were known to be operating passenger services over 13 Federal routes and cargo services over 17 Federal routes, and were operating a considerable, but unknown, number of local services.

However, as early as 1943 Sergei Ilyushin's design bureau began work on a more modern and faster aircraft intended as a replacement for the Li-2. The new aircraft was the Il-12, which made its first flight early in 1946 and was shown in public for the first time, on 18 August, 1946, at Tushino. Aeroflot began passenger operation with Il-12s on 22 August,

Aeroflot's Il-12 SSSR-L1403 at Vnukovo Airport, Moscow, in May 1948.

1947. Largescale operation of Il-12s may have started with the introduction of summer schedules on 25 May, 1948, and on that day SSSR-L1403 is known to have flown the inaugural Il-12 service over the Tashkent–Moscow route.

In layout the Il-12 was a low-wing cantilever monoplane with single fin and rudder, twin engines and fully-retractable nosewheel undercarriage. The engines were two 1,650/1,775 hp Shvetsov ASh-82FN fourteen-cylinder two-row air-cooled radials enclosed in close-fitting cowlings and driving four-blade constant-speed fully-feathering airscrews with fluid deicing. The ASh-82FN was a development of the M-82 which was a Soviet-built Pratt & Whitney R-1830.

The wing was a three-piece two-spar all-metal structure which tapered in chord and thickness but had an untapered stepped-forward leading edge to the centre section. Dihedral was 5 deg, there were split flaps, and production aircraft had thermal leading edge deicing.

The Aeroflot Il-12B SSSR-L1819 at Helsinki in April 1956. Markings were medium and dark blue. (*John Stroud.*)

The fuselage was an oval-section semi-monocoque structure and was unpressurized. The tail unit was of metal construction with fabric-covered control surfaces. The twin wheels of the main undercarriage retracted forward into the engine nacelles to be enclosed by two doors, and the single nosewheel retracted backward. Fuel was carried in eight tanks with a total capacity of 3,500 litres (769 Imp. gal). The Il-12's wings, engine cowlings and nose were flush riveted but the rest of the structure had mushroom-headed rivets.

Normal operating crew comprised two pilots and, aft of semi-bulkheads, navigator and radio operator, and there was also a jump-seat for a flight engineer. Apart from the two main forward flight deck windows there were five quite large windows in each side. There was a cargo and baggage compartment between the flight deck and the passenger cabin. The entrance door was aft on the starboard side and behind this was a lavatory, also on the starboard side. Beside and aft of the lavatory was a freight hold with a hatch in its port side, also accessible via a door in the rear wall of the cabin. A small galley could be installed on the port side opposite the entrance door. Some aircraft had luggage racks running the length of each side of the cabin and others had light fold-up racks. Most aircraft had eight rectangular double-glazed windows in each side of the passenger cabin and there was another rectangular window in each side of the radio/navigation cabin. The cabin was heated by hot air fed through ducts at the base of the walls. Ventilators and steward call bells were installed above each cabin window.

The number of seats installed in Aeroflot's Il-12s when they entered

The port engine of the Aeroflot Il-12 SSSR-L1701, seen at Malmi Airport, Helsinki, in 1948. The cowling and exhaust system should be compared with that of the Il-12B opposite.

service is not known but it is thought to have been 27 in nine rows, with double seats on the port side and single seats to starboard. In 1958 and 1959 Aeroflot's Il-12s were operated on international routes with 21 seats.

Aeroflot used Il-12s very widely on domestic services and in mid-1950 they were operating from Moscow to Vladivostok in 33 hr with nine stops en route. During the summer of 1955 the type was scheduled to operate the following international services: Moscow–Kazan'–Sverdlovsk–Omsk–Novosibirsk–Krasnoyarsk–Irkutsk–Ulan Bator–Sayn Shanda–Peking, Moscow–Kazan'–Sverdlovsk–Omsk–Novosibirsk–Krasnoyarsk–Irkutsk–Ulan Bator, Moscow–Vil'nyus–Prague, Moscow–Vil'nyus–Warsaw–Prague, Moscow–Kiev–Odessa–Bucharest–Sofia, Moscow–Kiev–L'vov–Budapest–Vienna, Moscow–Kiev–L'vov–Budapest–Belgrade–Tirana, Moscow–Vil'nyus–Warsaw–Berlin, Moscow–Leningrad–Helsinki and Moscow–Ural'sk–Aktyubinsk–Dzhusaly–Tashkent–Termez–Kabul.

In the winter 1955–56 the Il-12 schedule for the Moscow–Sverdlovsk–Novosibirsk–Irkutsk route was 17 hr 50 min with 14 hr 35 min flying time. The last known passenger operations by Aeroflot Il-12s were over the Zaporozh'ye–Khar'kov–Simferopol–Moscow route by the Ukrainian Directorate but these are believed to have ceased in May 1965.

Quite soon after their introduction some of Aeroflot's Il-12s were converted to freighters with large double doors on the port side. In 1962, Il-12 cargo aircraft are known to have been in service with at least four Aeroflot Directorates and operating scheduled cargo services over 10 Federal routes. It is not known whether the cargo Il-12s had a special designation but they may have been called the Il-12T.

The Il-12 seems to have suffered from a lack of directional control with one engine inoperative, and soon after it entered service a modified version was produced. This was the Il-12B with dorsal fin and aerodynamically-cleaner engine nacelles and exhaust system. It is thought that many of

This view of the Il-12B SSSR-L1819, at Helsinki, shows the cleaner nacelles of this version. (*John Stroud.*)

Aeroflot's Il-12B SSSR-L1723 photographed in 1956.

the original Il-12s were modified but known that not all were, and as late as 1957 cargo Il-12s were certainly flying with the original fin.

Polskie Linie Lotnicze bought five Il-12Bs, SP-LHA–SP-LHE, and the first of them was registered on 24 April, 1949. For some years they operated LOT's main services but all were withdrawn by 1959.

Československé Aerolinie (ČSA) bought several Il-12Bs and one of them, OK-CBA, was the first of the type to visit the United Kingdom when it replaced a DC-3 on the Prague–Northolt service on 12 September, 1949. That particular Il-12B had previously borne a Soviet registration and had accommodation for 28 passengers in seven rows of seats with double units each side of the central aisle. By moving the front bulkhead forward and reducing baggage and cargo space 32 seats could be installed.

In addition to Aeroflot, LOT and ČSA, the Soviet Air Force is known to have used Il-12s, and in May 1956 a Chinese-operated Il-12, with the number 5105 on its unmodified fin, made a flight from Peking to Lhasa in Tibet. It is possible that this aircraft was operated by the Chinese airline CAAC.

The number of Il-12s is not known but it is believed that Aeroflot had at least 200 by late 1949 and that some 120–150 were required for daily operation of scheduled services.

Československé Aerolinie's Il-12B OK-CBA. Markings were dark blue.

Two Soviet Air Force Il-14s at London Airport (Heathrow) in 1956. The nearest aircraft has the number 4340303 on its fin. (*Aeroplane*.)

A development of the Il-12 was the Il-14 which made its first flight in 1953, production aircraft beginning to appear in 1954. It is thought that first deliveries were made to the Soviet Air Force, and the first one seen outside the Soviet Union was the military example 4340203 which arrived at London Airport (Heathrow) on 23 February, 1955, carrying the Soviet Deputy Foreign Minister. The first civil version was the Il-14P (Passazhirskii–passenger) and this entered service with Aeroflot on 30 November, 1954.

The Il-14 was essentially an Il-12 with modifications. The engines were 1,630/1,900 hp Shvetsov ASh-82T fourteen-cylinder two-row air-cooled radials driving 3·8 m (12 ft $5\frac{3}{4}$ in) diameter AV-50 four-blade fully-feathering airscrews. The engine cowlings were considerably improved and the nacelles were carried right aft to the wing trailing edges. The exhaust was ejected via two thrust-augmentation tubes at the rear of each nacelle. The wing had three spars and was modified to provide greater lift. It was of modified TsAGI SR-5M section with thickness/chord ratio of 18 per cent at the roots and 12 per cent at the tips, compared with 16 per cent and 10 per cent for the Il-12. The increased centre-section chord of the Il-12 was eliminated and the wing tips were squarer. A new vertical fin and rudder was used and the aircraft was generally cleaned up.

The Aeroflot Il-14P SSSR-L1870.

Two cargo Ilyushins at Vnukovo Airport, Moscow, in about 1957. On the left is a cargo Il-12 with unmodified fin, and on the right an Il-14G or Il-14T. In the background is another Il-14.

The undercarriage was hydraulically operated, the main units having free fall and the nosewheel an emergency system. Main-wheel tyres had 5·2 kg/sq cm (73·96 lb/sq in) pressure and the nosewheel tyre 4·5 kg/sq cm (64 lb/sq in).

When it entered service with Aeroflot the Il-14P, which was referred to by the airline simply as the Il-14, had seats for 18 passengers, with six rows arranged in pairs on the port side and singly on the starboard. Forward of the passenger cabin was a buffet to starboard and a baggage hold to port. At the rear of the cabin on the port side was coat-hanging space and right aft were a lavatory and baggage hold.

In 1956 the Il-14M (Modifikatsyi—modified) appeared, this was 1 m (3 ft $3\frac{1}{4}$ in) longer and could have accommodation for 28–32 passengers, but Aeroflot installed only 24 seats in six rows, with double seats each side of the aisle. But in 1958 Aeroflot announced that it was increasing the Il-14P's seating to 24, and some time later Il-14M seating was increased to 32–36. In 1967 Il-14s were in service with the following seating capacities: 28, 32 and 36. Seat pitch was 870–1,000 mm (34·25–39·37 in) according to configuration.

Some Il-14Ps were lengthened and brought to Il-14M standard, and although it cannot be confirmed, there is some evidence to suggest that some Il-12s were converted to Il-14s, the same registrations having been seen on both types.

East German Deutsche Lufthansa's VEB Il-14P DM-SBU. This aircraft passed to Interflug.

There was also a cargo version of the short-fuselage Il-14 with double doors on the port side, generally known as the Il-14T; but there was a cargo version in service in 1960 with the designation Il-14G.

Soviet transports, in general, have always had less accommodation than their western counterparts and there is some evidence that they have not had their load-carrying abilities fully utilized. In 1960 this fact was realized by a staff member of Aeroflot, who put up a proposition for increasing the payload of the Il-14 series. Assumed structure weights were 12,420 kg (27,380 lb) for the Il-14P, 12,625 kg (27,832 lb) for the Il-14M and 12,290 kg (27,094 lb) for the Il-14G. Based on the maximum take-off weight of

LZ-ILD, one of Tabso's German-built VEB Il-14Ps, seen at Frankfurt-am-Main. (*Deutsche Lufthansa.*)

17,500 kg (38,581 lb) for all versions, and allowing for crew, fuel and oil and service equipment, this gave payloads of 3,105 kg (6,845 lb) for the Il-14P, 3,350 kg (7,385 lb) for the Il-14M and 3,600 kg (7,936 lb) for the Il-14G. Seven aircraft operated by the Kirgiz Directorate were weighed and found to average 1,825 kg (4,023 lb) less than their assumed weights and this meant that with minimum fuel and maximum payload these aircraft were being underloaded by about 2 tons on each flight. It was found that the Il-14P with 24 seats could carry an extra 2,000 kg (4,409 lb) or with 32 seats an extra 1,500 kg (3,306 lb). Full details of the results of this study are not known but during 1965 at least one version of the Il-14 had its cargo capacity increased to 2,396 kg (5,282 lb).

Soviet figures for cabin dimensions and doors of the passenger aircraft are: passenger cabin length 8·9 m (29 ft 2¼ in), width 2·67 m (8 ft 9 in), height 1·94 m (6 ft 4½ in) and volume 35·3 cu m (1,246·6 cu ft); front hold volume 4 cu m (141·26 cu ft), rear hold volume 4·4 cu m (155·38 cu ft); passenger door size 1·44 m (4 ft 8½ in) by 68 cm (2 ft 2¾ in), front hold door 95 cm (3 ft 1·4 in) by 58 cm (1 ft 10·8 in) and rear hold door 70 cm (2 ft 3½ in) by 60 cm (1 ft 11½ in). Passenger door sill height is 2·45 m (8 ft 0¼ in) and the front and rear hold sill heights are 2·75 m (9 ft 0¼ in) and 2·65 m (8 ft 8¼ in) respectively.

Aeroflot used very large numbers of Il-14s on Federal, international and

The Avia-14 Series

These photographs show the six Czechoslovak-built versions of the Ilyushin Il-14. Only the Super and the Salon are significantly different to the Soviet-built aircraft.

SP-LNN, the sole Avia-14 in the Polskie Linie Lotnicze Il-14 fleet. It is seen in LOT's latest blue and white colour scheme. (*Polskie Linie Lotnicze.*)

OK-LCA, the Avia-14P before delivery to ČSA. (*Aeroplane.*)

3X-BKE, Air Guinée's Avia-14T cargo aircraft.

ČSA's Avia-14-32A OK-MCL *Trenčin*. Markings are red. (*ČSA*.)

Avia-14 Super operated by Tabso. This version has a pressurized cabin, circular windows and removable wing tip fuel tanks. The fuselage stripe and tail markings are light blue and the Bulgarian flag is white, green and red.

The Avia-14 Salon seen with wing tip fuel tanks.

local services. As late as the summer of 1967 they were being operated on Federal routes by 27 Directorates and on local services by 26 Directorates. In 1966 they were operating passenger services on at least 80 Federal routes and cargo services over nine Federal routes; the number of local services still being operated at that time by Il-14s is unknown but must have been considerable. The last known introduction of Il-14s on an Aeroflot route was on 28 February, 1962, when Il-14Ms began operating between Syktyvkar and Sverdlovsk.

Apart from the largescale Soviet production of Il-14s, the type was built under licence at Dresden by VEB Flugzeugwerke and in Czechoslovakia by Avia.

The Dresden-built aircraft were Il-14Ps and bore the designation VEB Il-14P. The first of these was DDR-AVF and it was reported as making its first flight in April 1956. German production is believed to have been between 60 and 80 aircraft.

One of the CAAC Il-14s at Peking in March 1959.

Well over 50 Il-14s are reported to have been built by Avia. The first examples were completed in 1957 and initial production aircraft bore the designations Avia-14 and Avia-14P. Later that year came the 32-passenger Avia-14-32A, equivalent to the Il-14M, and the cargo Avia-14T. In 1960 the Avia-14 Super and Avia-14 Salon were produced. The Salon was an executive version of the Avia-14-32A, but the Super was pressurized, had circular cabin windows and seats for up to 42 passengers. Both the Salon and the Super could be fitted with long-range wing tip fuel tanks.

Total production of Il-12s and Il-14s is not known but figures of as high as 3,000 Il-12s and about 3,500 Il-14s have been reported.

The Il-14 was the first Soviet transport aircraft to be exported in any quantity. Air Guinée had six including an Avia-14T, two Avia-14-32As and an Avia-14 Super, and also leased from ČSA the Avia-14-32A OK-MCO which was operated in Air Guinée livery but with Czechoslovak registration; Air Mali had three Avia-14-32As; Mongolian Airlines operated Il-14s; CAAC in China had Il-14Ps and Il-14Ms, possibly built in China; ČSA had no less than 26, comprising six Avia-14Ps, 19 Avia-14-32As and

A Malév Il-14 taking-off from Stockholm.

an Avia-14T; Cubana had 12; Deutsche Lufthansa (DDR) and Interflug had 39, all VEB Il-14Ps except for one Soviet-built Il-14P; JAT had six; Malév had 10 in a mixed fleet of Il-14Ms and VEB Il-14Ps; Polskie Linie Lotnicze had 16 comprising six Il-14Ps, nine VEB Il-14Ps and an Avia-14; Tabso had 11 which included the variants Il-14P, Il-14M, VEB Il-14P and Avia-14 Super; Tarom had 18, mostly VEB Il-14Ps but with some Il-14Ps and Il-14Ms; Ukamps (North Korea) had Il-14s; Yemen Airlines had at least one Il-14P; and United Arab Airlines used Il-14Ps.

The air forces of the USSR, Czechoslovakia, Poland, Albania, India and the United Arab Republic are among those known to have employed Il-14s; and many have been presented to Heads of States, some of the recipients being Prime Minister Nehru, President Sukarno, President Nasser, the Shah of Iran and the Prime Minister of Burma whose aircraft was appropriately registered XY-VIP.

ČSA's first service flight with an Il-14 was by OK-LCB on 8 June, 1957. LOT has no record of its first Il-14 service but it is known that the first

Jugoslovenski Aerotransport's Il-14 YU-ADG at Frankfurt-am-Main Airport. This aircraft was almost certainly an Il-14P. (*Deutsche Presse-Agentur GmbH.*)

aircraft, SP-LNA, was registered on 20 June, 1955. Deutsche Lufthansa (DDR) began operations in September 1955 when the Soviet-built Il-14P DDR-ABA made its first Berlin–Moscow flight, and East German domestic services appear to have started on 1 May, 1956, when an Il-14 opened the Dresden–Karl-Marx-Stadt (Chemnitz) route. Dates of Il-14 introduction by other airlines are not known.

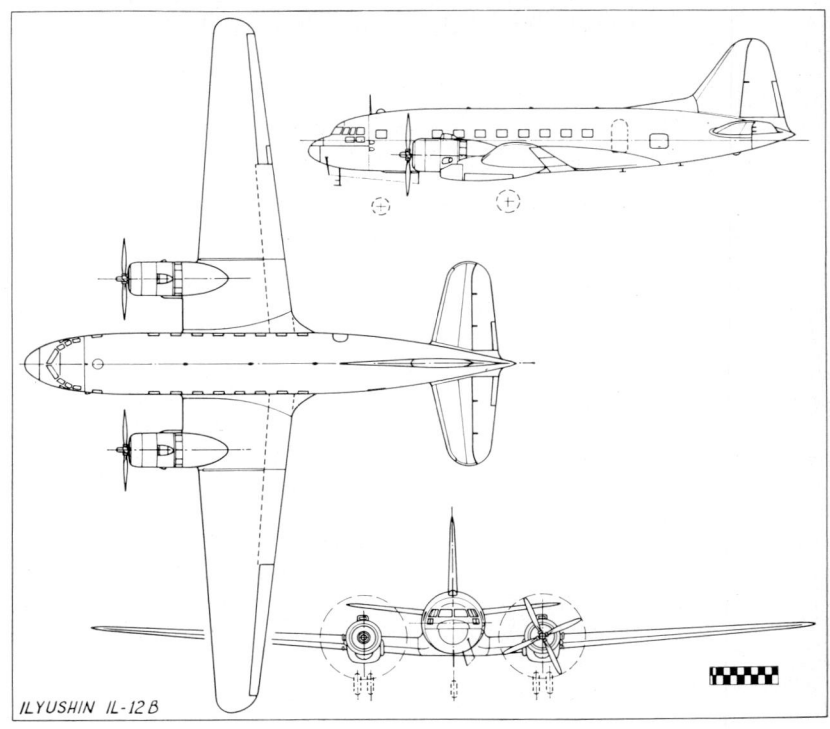

ILYUSHIN IL-12B

As late as 1966 Aviaexport stated that five versions of the Il-14 were available for export. They were the Il-14P (28–32 passengers, baggage, mail and cargo), the Il-14M (32–36 passengers, baggage, mail and cargo), the Il-14T cargo aircraft with 3,000 kg (6,613 lb) payload, a 10–18 seat de luxe version and a photographic survey version.

Constructor's numbers of the Il-14 series are quite complicated, and attempts have been made to use them to assess total production; this exercise misleads because the reasons for the complete numbering system are not known. In 1954 aircraft bore c/ns in the 4340000 series, examples being known between 4340203 and 4340607; 1955 c/ns were in the 5440000 series with only 5440709 being known. At the start of 1956 the 6340000 series was in use, with examples identified between 6341102 and 6341706.

During 1956 a new system was introduced consisting of nine figures, the earliest known being 146000401. The 14 stood for Il-14, the 6 for 1956 and the last four digits were the actual aircraft number. In 1957 the third figure was changed to 7, and in the following year became 8 to show 1958 production. From the c/ns traced it is thought that each batch in the new numbering system consisted of 50 aircraft.

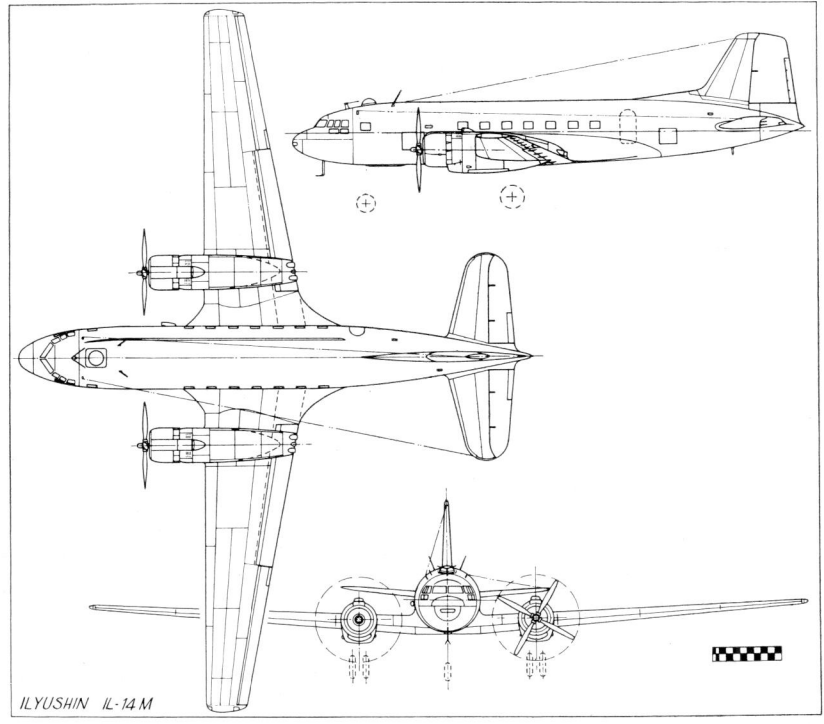

ILYUSHIN IL-14M

The Dresden-built aircraft had c/ns following on from the Soviet system and were in the 14803000 series, 31 c/ns being known between 14803010 and 14803079.

Several systems seem to have been used for the Avia-14s. Serials in the 703000 and 705000 groups were used for aircraft built in 1957; 805000, 806000 and 807000 in 1958; and 013000 and 014000 in 1960. The Avia-14-32A TZ-ABH of Air Mali has c/n 7342501, which does not appear to fit into any system, but their Avia-14-32As TZ-ABF and TZ-ABG (c/ns 147001310 and 146001050) are almost certainly Soviet-built aircraft which were modified in Czechoslovakia.

Known scheduled services by Il-14s operated by non-Soviet airlines in the summer of 1967 appear in Appendix IV.

The cabin of a ČSA Il-14. Looking aft (left) and forward (right). (*ČSA*.)

Il-12B

Span 31·7 m (104 ft); length 21·31 m (69 ft 11 in); height 8·07 m (26 ft 5¾ in); wing area 100 sq m (1,076·39 sq ft).

Empty weight 9,000 kg (19,841 lb); payload 3,000 kg (6,613 lb); maximum take-off weight 17,250 kg (38,030 lb).

Maximum speed at 2,500 m (8,202 ft) 407 km/h (219·61 kt) (252·9 mph); cruising speed at 2,500 m (8,202 ft) 350 km/h (188·86 kt) (217·48 mph); landing speed 145 km/h (78·24 kt) (90·1 mph); take-off run at maximum weight 520 m (1,706 ft); landing run 450 m (1,476 ft); service ceiling 6,700 m (21,981 ft); single-engine ceiling 3,000 m (9,842 ft); range with 27 passengers 2,000 km (1,079 n.miles); range with 32 passengers 1,250 km (674 n.miles).

Weights are for 27-passenger version.

Il-14

Span 31·7 m (104 ft); length 21·31 m (69 ft 11 in) Il-14P and Il-14T, and 22·31 m (73 ft 2¼ in) Il-14M; height 7·8 m (25 ft 7 in); aspect ratio 10; wing area 100 sq m (1,076·39 sq ft); track 7·7 m (25 ft 3 in); wheelbase 5·37 m (17 ft 7¼ in).

Empty weight 12,500 kg (27,557 lb) Il-14P, 12,700 kg (27,998 lb) Il-14M and 12,880 kg (28,395 lb) Il-14T; maximum payload 3,000 kg (6,613 lb) Il-14P, 3,300 kg (7,275 lb) Il-14M and 3,100 kg (6,834 lb) Il-14T; maximum fuel (normal tanks) 2,580 kg (5,688 lb); maximum fuel with supplementary tanks 3,180 kg (7,010 lb); maximum take-off weight (all versions) 17,500 kg (38,581 lb); maximum landing weight (emergency only) 17,000 kg (37,479 lb) Il-14P, and 17,250 kg (38,030 lb) Il-14M and Il-14T.

Maximum speed at 2,400 m (7,874 ft) 431 km/h (232·56 kt) (267·81 mph); maximum cruising speed at 2,500–3,000 m (8,202–9,842 ft) 350 km/h (188·86 kt) (217·48 mph); economic cruising speed at 2,500–3,000 m (8,202–9,842 ft) 320 km/h (172·67 kt) (198·84 mph); minimum speed 110 km/h (59·36 kt) (68·35 mph); take-off speed, with 20 deg flap, 165 km/h (89·03 kt) (102·52 mph); landing speed, with 45 deg flap, 135 km/h (72·84 kt) (83·88 mph); landing speed, without use of flaps, 145 km/h (78·24 kt) (90·09 mph).

Take-off run on paved surface 485 m (1,591 ft); take-off distance to 15 m (49 ft) 900 m

(2,952 ft); landing run at 15,880 kg (35,009 lb) 443 m (1,453 ft); landing distance from 15 m (49 ft) 1,000 m (3,280 ft).

Service ceiling 7,400 m (24,278 ft).

Range (Il-14M) at maximum take-off weight at 2,500–3,000 m (8,202–9,842 ft) with 3,300 kg (7,275 lb) payload and 1 hr fuel reserve, 400 km (215 n.miles); range under same conditions with 2,300 kg (5,070 lb) payload 1,200 km (647 n.miles), and with 1,600 kg (3,527 lb) payload 1,750 km (944 n.miles).

Fuel consumption at 2,500–3,000 m (8,202–9,842 ft) and 320 km/h (172·67 kt) (198·84 mph) 200 kg (440 lb) per engine per hour.

Data from Soviet sources except minimum speed, take-off distance and landing distance which are for the Czechoslovak Avia-14-32A.

The 60-passenger piston-engined Il-18 of 1947.
(*Courtesy Aviation Magazine International.*)

Ilyushin Il-18

The original Ilyushin Il-18 was a completely different aircraft to the Il-18 which is today well known as the Soviet airscrew-turbine powered transport which entered service with Aeroflot in 1959 and has since become one of the most widely used Soviet transport aircraft.

This first Il-18 is believed to have been designed to meet an Aeroflot specification, and made its first flight on 30 July, 1947. It was a low-wing cantilever monoplane with single fin and rudder, four piston engines and retractable nosewheel undercarriage.

The wing had dihedral and marked taper on leading and trailing edges. The fuselage was of circular section and had accommodation for 60 passengers, six crew and 900 kg (1,984 lb) of baggage. The nose was completely faired into the fuselage without a break at the flight deck windows. The fin and rudder were of high aspect ratio, finely tapered, and there was a small dorsal fin. The rudder appears to have been fabric-covered. All undercarriage units had twin wheels, the nosewheels retracted backward into the fuselage and the main units retracted forward into the inboard engine nacelles. It has been reported that the Il-18 was unpressurized but presumably production aircraft would have been pressurized.

The four engines were 2,300 hp Shvetsov ASh-73 eighteen-cylinder air-cooled radials and drove four-blade variable-pitch airscrews.

Flying for March 1949 reported that the Il-18 had recently gone into service on the Moscow–Khabarovsk route and would be the main type to operate between Moscow and Vladivostok, and a schedule of unknown origin for departures from Vnukovo Airport, Moscow, during the summer of 1950 showed the following Il-18 operations: Moscow–Sverdlovsk–Novosibirsk–Kirensk–Yakutsk, 10 times a month; Moscow–Sverdlovsk–Omsk–Krasnoyarsk, daily; Moscow–Sverdlovsk–Novosibirsk, daily; Moscow – Sverdlovsk – Novosibirsk – Krasnoyarsk – Chita – Tygda–Khabarovsk, daily; Moscow–Sverdlovsk–Novosibirsk–Krasnoyarsk–Irkutsk, daily; Moscow–Sverdlovsk–Omsk–Krasnoyarsk–Irkutsk–Tygda–Khabarovsk–Vladivostok, daily; and Moscow–Sverdlovsk–Novosibirsk–Krasnoyarsk–Irkutsk–Tygda–Khabarovsk, daily. In addition, the Il-18 was shown along with Li-2s and Il-12s as operating: Moscow–Rostov–Sukhumi–Yerevan, daily; Moscow–Khar'kov–Sukhumi–Kutaisi–Tbilisi, daily; Moscow–Stalino (Donetsk)–Kutaisi, daily; Moscow–Sukhumi–Tbilisi, daily; Moscow–Khar'kov–Rostov–Krasnodar, daily; and Moscow–Stalino–Sukhumi, daily. Also at that time a schedule showed a daily Il-18 service from Moscow's Lyubertsy Airport to Aktyubinsk and Tashkent.

Because traffic volume did not require numbers of such a large aircraft, Aeroflot decided against the Il-18, flight trials ended in 1948 and the aircraft was not put into production.

The source of *Flying*'s report is not known; neither is there an explanation for the 1950 schedules showing Il-18s operating regularly from Moscow.

Dimensions are not known. Empty weight 28,490 kg (62,809 lb); loaded weight 42,500 kg (93,696 lb). Performance has been reported as: maximum speed 588 km/h (317·28 kt) (365·37 mph); maximum cruising speed 510 km/h (275·2 kt) (316·9 mph); time to 5,000 m (16,404 ft) 13 min 9 sec; service ceiling 10,700 m (35,104 ft); and maximum range 6,200 km (3,345 n.miles).

The prototype airscrew-turbine Il-18 Moskva (Moscow), SSSR-L5811, flying over Moscow in 1957.

Ilyushin Il-18

In 1953 Aeroflot began work on a major re-equipment programme, the initial result of which was the twin-jet Tupolev Tu-104 which first flew in 1955 and entered service in September 1956. Even with production of the Tu-104, Aeroflot was still left with an urgent need for high-capacity aircraft capable of operating over a high percentage of its route system and having the ability to operate from modest aerodromes. Two aircraft were designed to meet these requirements—the Antonov An-10 and the Ilyushin Il-18.

The Il-18 design would appear to have started in 1954 or early 1955, and the decision to build was taken early in 1956. The new aircraft was a pressurized medium-range type powered by four propeller-turbines and was initially planned for 75–100 passengers. It was a completely new design; it was not adapted from a military aircraft and was in no way a direct development of the piston-engined Il-18 of 1947.

The prototype Il-18, SSSR-L5811, first flew on 4 July, 1957, and was seen in public on 10 July when it was shown at Vnukovo Airport, Moscow, in company with the prototype An-10 and the Tu-110, both of which were also making their first appearance. The Il-18 bore the name MOCKBA (Moscow) in large letters above the cabin windows, and for some time this name was used as a type designation but in service the aircraft has always been known simply as the Il-18.

A batch of 20 pre-production aircraft was laid down, some with 4,000 ehp Kuznetsov NK-4 engines and some with the 4,000 ehp Ivchenko AI-20, and it was the latter which was chosen for production aircraft. FAI records set by Il-18s in November 1958 show the aircraft concerned as having AI-20 and TV-20 engines but these are thought to be different designations for the same engine. Of these early Il-18s only SSSR-L5818,

5819, 5820 and 5821 are known. The Soviet Air Force had some Il-18s by 1958 and it is possible that some were from the pre-production batch. Production of the Il-18 began at the end of 1957 and, after a period of intensive route trials including operation of cargo services from Moscow to Adler, Baku and Frunze, the type entered service with Aeroflot on 20 April, 1959, on the Moscow–Alma Ata and Moscow–Adler/Sochi routes.

Aeroflot Il-18s are registered in the SSSR-75000 series but these numbers were not allocated in strict sequence. The first batches were in the 75600 series and ran to at least 75894, the highest number known. Then came batches in the 75500 series and, more recently, aircraft running from 75400. Polar Directorate aircraft have registrations beginning with 0, such as

An early Aeroflot Il-18 at Vnukovo Airport, Moscow, in 1958, the year before the type went into passenger service. The aircraft in the background include six Tu-104s.

SSSR-04330. Constructor's numbers comprise nine figures and were allocated in groups of five. A complete number is 189001503, for SSSR-75702. The 18 represents the type number and the 9 almost certainly shows the aircraft as having been built in 1959. The 15 is the batch number and 03 is the 3rd aircraft in the 15th batch. Examples of actual aircraft numbers in sequence are 4001–5, 4101–5, 4201–5, and thus it is likely, although it cannot be confirmed, that 189001503 was the 73rd production Il-18. After reaching the 9000 series, the fifth digit also became part of the aircraft c/n. The highest known c/n is that of Aeroflot's SSSR-75452, 187010102.

Production of Il-18s appears to have been about three a month up to the end of 1960, perhaps double that rate in 1961 and 1962, and gradually declining thereafter. Deliveries of the latest version are known to have been made during the summer of 1967 and total production is likely to have exceeded 500 by that time, 216 Soviet-registered aircraft having been identified.

The Il-18 entered Aeroflot service as an 80-seat aircraft with accommodation in two cabins. The main cabin, with 13 windows on each side, had accommodation for 65 with 13 rows of seats arranged in triple units on the starboard side and pairs on the port side. Forward of the galley and cloakroom area was a cabin for 15, with 11 seats on the starboard side and four on the port side immediately aft of the entrance door. In this layout there was one window on the port side of the front cabin and three on the

SSSR-75768, an early Aeroflot Il-18, on final approach to London Airport (Heathrow) on 14 October, 1959, when the type temporarily took over operation of the Moscow–Copenhagen–London service from Tu-104As. This was the first service to be operated to London by a large airscrew-turbine transport of non-British design and manufacture. Because of bad weather at Copenhagen, SSSR-75768 had on that day flown via Amsterdam.

starboard side. ČSA's first Il-18, OK-NAA, was of similar configuration but had only 12 main cabin windows on the port side.

Later Il-18s had re-arranged interiors and the window arrangement was altered. The forward door was located further aft, and in a cabin forward of this door were three windows on each side. The number of windows on each side of the main cabin was reduced to nine although some aircraft have a 10th window on the starboard side. The rear door was also re-positioned and a cabin with three windows a side was located aft of this door. Operated in 73–84 and 89–111 seat configurations, this version would seem to be the Il-18V, frequently misreported as the Il-18B. It appears to have entered service in 1962 and so far the Il-18V seems to have been the most used variant. A cut-away drawing of an Aeroflot Il-18, published soon after the type entered service, showed it with 55 tourist class seats in the main cabin, 20 tourist class seats in the forward cabin and eight first class seats in the aft cabin. In most Aeroflot aircraft there is accommodation for more passengers in summer than in winter because of the greater space required in winter for stowage of passengers' heavy clothing. Normal operating crew of the Il-18 is five.

Aeroflot's Il-18 SSSR-75768 taxi-ing in at London Airport (Heathrow) on 14 October, 1959. This example is one of the early aircraft with forward-positioned entrance and single forward cabin window. (*British European Airways.*)

The Air Mali Il-18 TZ-ABD. The fuselage stripe is yellow with green outline. The Mali flag is green, yellow and red with green forward. Lettering is black. (*Courtesy John W. R. Taylor.*)

In the summer of 1962 Il-18s were in service with Aeroflot in 80, 84, 88 and 89 seat configurations, were used by 13 Directorates and Groups and operating over more than 70 Federal routes; in the winter of 1965–66 Il-18s were flying in 79, 87, 89 and 110 seat configurations and serving with 14 Directorates; in 1966 there were 16 Aeroflot Directorates and also 235 Division using Il-18s, the type was working on more than 180 routes and during the year carried $10\frac{1}{2}$ mn passengers. There are probably more Il-18s in service with Aeroflot than any other major type, and in 1967 it was reported that they were carrying 40 per cent of all Aeroflot's traffic.

In addition to the very large number of Il-18s supplied to Aeroflot, the type was supplied to Air Guinée, Air Mali, CAAC (China), ČSA, Cubana, Deutsche Lufthansa (later Interflug), Ghana Airways, Malév, Polskie Linie Lotnicze, Tabso and Tarom. The first non-Soviet operation is believed to have been by ČSA, which flew its first Il-18 service, with OK-NAA, on 28 January, 1960, between Prague and Bratislava. Non-airline Il-18s have been supplied to a number of countries including Algeria, the German Democratic Republic and Poland, the last two using them in their air forces. At least 84 Il-18s have been supplied to non-Soviet airlines and the last order to be announced, in 1967, was for four aircraft for United Arab Airlines. Some of the exported aircraft are known to have served previously with Aeroflot.

ČSA's Il-18V OK-OAC *Sliač*. On the right is the Tu-104A OK-LDA *Praha* (*Prague*). (*ČSA.*)

An Il-18 flight of which the Russians are particularly proud was that made in December 1961 by SSSR-75743 from Moscow to Mirnyy, in the Antarctic, via Tashkent, Delhi, Djakarta, Sydney and Christchurch. The journey of 25,793 km (13,918 n.miles) was flown in 44 hr 34 min.

In layout the Il-18 is an elegant low-wing cantilever monoplane with single fin and rudder and fully-retractable undercarriage. The wing is an all-metal structure having an aspect ratio of 10 and a mean thickness/chord ratio of 14 per cent. In the Il-18 and Il-18V it is thought that there is a short centre section to which is bolted three-spar inner wings

Ghana Airways' Il-18V 9G-AAN at Lagos, Nigeria, on 14 March, 1964.
(*John Stroud.*)

which carry the engine mountings. To these are attached the outer two-spar wing sections. The wing has dihedral and tapers in chord and thickness. The ailerons are mass-balanced and aerodynamically-compensated and the starboard aileron incorporates a trim tab. All control surfaces are manually-operated and without power assistance. Single-slotted flaps extend from the fuselage to the ailerons (take-off flap setting is 15–30 deg and landing setting 30 deg). Electrical hot-element leading edge deicing is provided. The inner sections of the wing each contain 10 flexible bag tanks and the outer wing spars form integral fuel tanks with a total capacity of 23,700 litres (5,213 Imp. gal).

The fuselage is a circular-section semi-monocoque structure with closely spaced frames, and doublers round all cut-outs. Apart from the tail area the entire fuselage is pressurized to a maximum differential of 0·5 kg/sq cm (7·1 lb/sq in).

The tail unit is of conventional layout and all leading edges are protected from ice accretion by electrically-heated elements.

The undercarriage comprises four-wheel bogy main units and twin steerable nosewheels. All units are hydraulically-retracted forward and have free-fall capability in event of hydraulic failure. Braking and nosewheel steering is hydraulic with a pneumatic emergency system. Main-wheel tyre pressure is 8 kg/sq cm (113·79 lb/sq in) and nosewheel tyre pressure 6 kg/sq cm (85·34 lb/sq in). Except in the case of the earliest aircraft, the undercarriage doors are only open when the undercarriage is being retracted or lowered.

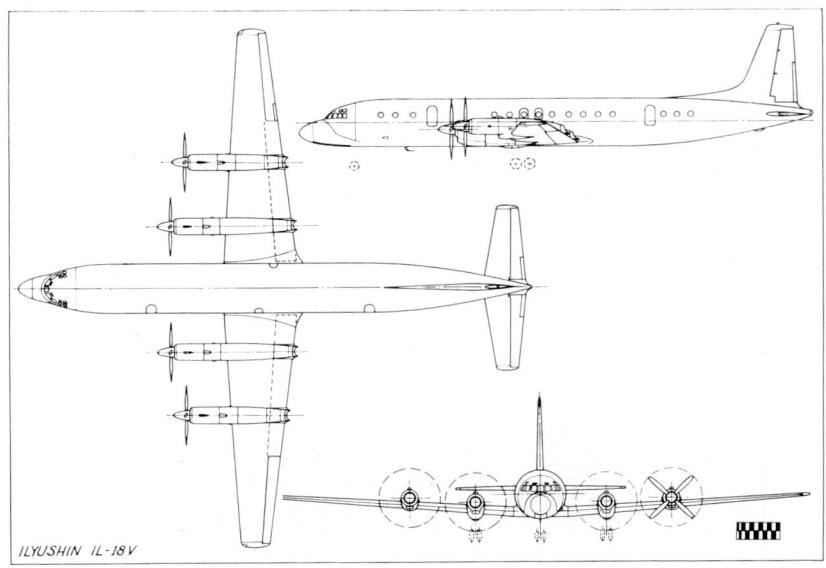

ILYUSHIN IL-18V

The engines in the Il-18V are four Ivchenko AI-20K or more recently installed AI-20M axial-flow turbines with 10-stage compressor and three-stage turbine. Take-off power of the AI-20K is 4,000 ehp and of the AI-20M 4,250 ehp. The engines drive AV-68I automatically-feathering four-blade airscrews of 4·5 m (14 ft 9 in) diameter. Airscrew blades, spinners and engine intakes have electro-thermal ice protection. The AI-20 is a

The main cabin of a Malév Il-18V. (*Flight International.*)

constant-speed engine with power being controlled by fuel-flow and airscrew-pitch, an automatic control system keeping the maximum rpm, thrust and gas temperatures within the allowable limits. The cowlings are of extremely clean design. Although the AI-20 had a bad reputation initially, particularly with non-Soviet airlines, by early 1967 its time between overhauls was 4,000 hr, and some have run to 6,000 hr.

The Il-18 airframe has a design life of 25,000 hr with a 6,000 hr overhaul life.

Standard configurations for the Il-18V are for 84–89 or 110 passengers, but there is an executive layout for 35–50 passengers. The passenger cabin is 24·36 m (79 ft 11 in) long, 3·23 m (10 ft $7\frac{1}{2}$ in) wide and 2 m (6 ft $6\frac{3}{4}$ in) high, with an overall volume of 104·57 cu m (3,692·85 cu ft). There

Polskie Linie Lotnicze's Il-18V SP-LSF on final approach to London Airport (Heathrow) on 5 September, 1967. (*John Stroud.*)

are underfloor holds fore and aft and these have volumes of 13·32 and 13·68 cu m (470·39 and 483·1 cu ft) respectively. An unpressurized baggage hold aft of the passenger area has a volume of 7 cu m (249·3 cu ft). The inward-opening passenger doors measure 76 cm by 1·4 m (2 ft 6 in by 4 ft 7 in) and the hold doors are 75 cm by 1·25 m (2 ft $5\frac{1}{2}$ in by 4 ft $1\frac{1}{4}$ in). Passenger door sill height is 3·1 m (10 ft 2 in). Two overwing emergency exits on each side measure 75 cm by 45 cm (2 ft $5\frac{1}{2}$ in by 1 ft $5\frac{3}{4}$ in).

In 1962 the Il-18I was announced as a development of the Il-18V with increased range, modified fuel system, increased take-off weight and greater cabin volume. This aircraft did not go into service until the end of 1965 when it appeared as the Il-18D (Dal'niy—long-range). During 1965 details were released of the Il-18D and Il-18E, but it was not until 1967 that it was known that Il-18I had only been a provisional designation for the Il-18D.

SSSR-75581 was described as an Il-18D at the Paris Aero Show in 1965. Fuselage markings on most Aeroflot Il-18s are blue but some with a mauve stripe and mauve fin and rudder have been seen in Moscow. (*John Stroud.*)

The Il-18E, like the Il-18D, has had the pressurized area of the fuselage increased by the removal of the aft pressure bulkhead from frame 56 to frame 62, a modification undertaken to meet seasonal traffic demands. Both the Il-18E and Il-18D are operated in four configurations. There is an all-tourist layout for winter operation with 90 seats arranged in triple units on the starboard side and pairs on the port side, the front cabin having four rows, the main cabin 11 rows and the tapering aft cabin three rows. The large coat storage area between the front and main cabins is removed in the summer to increase capacity to 100, with 13 rows of seats in the main cabin. There is a 110-seat mixed tourist–economy class with four rows of six-across seats in the front cabin, 11 rows of six-across and one of five-across in the main cabin, and three rows of five-across in the rear cabin. Removal of the coat space in this version allows installation of two more seat rows to provide accommodation for 122 in summer. Details have also been published of a 65-seat version with a mixture of double and triple seats to accommodate 57 in the front and main cabins and eight de luxe seats in the aft cabin. This last version is not known to have gone into airline service.

The Il-18E was already in production at the end of 1965 but the first

ČSA's Il-18D OK-WAJ. (*ČSA.*)

Il-18Es are believed to have been converted Il-18Vs put into operation at the beginning of 1965 by the Latvian Aviation Group, now a Directorate, and the Moscow Transport Directorate.

Cabin air-conditioning has been improved in the Il-18D and E, navigational aids have also been improved and ILS/VOR installed. The Il-18D is reported to have been fitted with the Polosa autoland system to meet ICAO Category III conditions (200 m (700 ft) Runway Visual Range Cat. IIIA and 45 m (150 ft) RVR Cat. IIIB).

The seats in the Il-18D and E are of a new design with increased comfort. Tourist class seats are 438 mm (17·24 in) wide and economy 410 mm (16·14 in) between arm rests and this is said to be acceptable for flights of $1\frac{1}{2}$–2 hr, but on longer stages passengers have complained of discomfort and have

The flight deck of the Aeroflot Il-18D SSSR-75581. (*Flight International.*)

criticized the fact that an aircraft which began service with 80 seats now has as many as 122. Seat pitch is 750 mm (29·5 in) in all-tourist and tourist–economy configurations.

Apart from its increased interior volume the Il-18D differs from the earlier Il-18s in a number of ways. The wing structure was changed to comprise a centre section which passes under the fuselage and carries all four engines. To this are fitted detachable outer wings. Centre-section fuel tanks provide additional capacity for 6,500 litres (1,430 Imp. gal) and the AI-20M engines have 10 per cent lower consumption than the AI-20K engines used in most of the earlier Il-18s. The increased tankage and improved consumption extend the range from the 4,800 km (2,590 n.miles) of the Il-18V and Il-18E to 6,500 km (3,507 n.miles). Payload of all versions is 13,500 kg (29,762 lb), but the Il-18D's take-off weight has been increased by 2,800 kg (6,172 lb) to 64,000 kg (141,096 lb). Navigational equipment was further improved and now includes Doppler, and the deicing system has been revised. Another change was the repositioning of the auxiliary power unit from the rear baggage hold to a position under the forward fuselage. The APU is covered by a shallow streamlined fairing and this is the main external feature by which the Il-18D can be recognized. The entire APU can be lowered for inspection.

While still retaining the original designation Il-18I, the aircraft made test flights over the Moscow–Irkutsk, Moscow–Leningrad, Moscow–Vladivostok, Khabarovsk–Tashkent and Tashkent–Arkhangel'sk–Moscow routes. It is not known why the aircraft was delayed in going into service but it only completed its State tests in the second half of 1965. At the end of that year Il-18Ds entered service with non-Soviet airlines and in 1966 Aeroflot put the type into regular operation.

The Il-18 series has performed magnificent service for Aeroflot and as late as 1967 formed the backbone of the fleet, it has also proved to be the most successful large Soviet aeroplane in overseas markets, with more than 80 going into service with airlines outside the Soviet Union.

Span 37·4 m (122 ft 8½ in); length 35·9 m (117 ft 9½ in); height 10·16 m (33 ft 4 in); wing area 140 sq m (1,506·95 sq ft); aspect ratio 10; track 9 m (29 ft 6¼ in); wheelbase 12·75 m (41 ft 10 in).

Il-18V

Empty weight 31,500 kg (69,445 lb); operating empty weight 34,500 kg (76,059 lb); maximum fuel 18,600 kg (41,005 lb); maximum payload 13,500 (29,762 lb); maximum take-off weight 61,200 kg (134,922 lb); maximum landing weight 51,200 kg (112,876 lb).

Maximum speed at 8,000 m (26,246 ft) 685 km/h (369·62 kt) (425·63 mph); cruising speed at 8,000 m (26,246 ft) 625–650 km/h (337·25–350·74 kt) (388·35–403·89 mph); take-off speed at 61,000 kg (134,482 lb) 220 km/h (118·71 kt) (136·7 mph); landing speed at 50,000 kg (110,231 lb) 200 km/h (107·92 kt) (124·27 mph).

Take-off run at 61,000 kg (134,482 lb) at sea level ISA 1,200 m (3,937 ft), at ISA plus 25 deg C. 1,425 m (4,675 ft), at 1,500 m (4,921 ft) elevation ISA 1,750 m (5,741 ft); take-off distance to 15 m (49 ft) at 61,000 kg (134,482 lb) at sea level ISA 2,050 m (6,725 ft), at ISA plus 25 deg C. 2,450 m (8,038 ft), at 1,500 m (4,921 ft) elevation ISA 3,050 m (10,006 ft); runway length required at 61,000 kg (134,482 lb) at sea level ISA 1,810 m (5,938 ft).

Landing run at 46,000 kg (101,413 lb) at sea level ISA, 720 m (2,362 ft).

Service ceiling at 61,000 kg (134,482 lb) 10,750 m (35,268 ft); three-engine service ceiling at 61,000 kg (134,482 lb) 6,500 m (21,325 ft).

Maximum payload range at 8,000 m (26,246 ft) with 1 hr fuel reserve 2,500 km (1,349 n.miles); maximum fuel range at 8,000 m (26,246 ft) with 9,000 kg (19,841 lb) payload and 1 hr fuel reserve, 4,800 km (2,590 n.miles).

Il-18E

Operating empty weight 34,630 kg (76,345 lb); maximum fuel 18,600 kg 41,005 lb); maximum payload 13,500 kg (29,762 lb); maximum take-off weight 61,200 kg (134,922 lb).

Performance as Il-18V.

Il-18D

Operating empty weight 35,000 kg (77,162 lb); maximum fuel 23,550 kg (51,918 lb); maximum payload 13,500 kg (29,762 lb); maximum take-off weight 64,000 kg (141,096 lb).

Cruising speed at 8,000 m (26,246 ft) 625 km/h (337·25 kt) (388·35 mph); take-off run at 64,000 kg (141,096 lb) at sea level ISA 1,450 m (4,757 ft); landing run 850 m (2,788 ft); cruise altitude 8,000 m (26,246 ft); maximum payload range at 610 km/h (329·16 kt) (379 mph) at 8,000 m (26,246 ft) in ISA with 1 hr fuel reserve, 4,000 km (2,158 n.miles); range with 10,300 kg (22,707 lb) payload and 1 hr fuel reserve, 5,000 km (2,698 n.miles); maximum fuel range with 6,500 kg (14,330 lb) payload and 1 hr fuel reserve, 6,500 km (3,507 n.miles).

The Polskie Linie Lotnicze Il-18V SP-LSG which was delivered in 1965. (*Polskie Linie Lotnicze.*)

Known dates for Il-18 introductions and technical proving flights:

Aeroflot

1959 April 20	Moscow–Alma Ata and Moscow–Adler/Sochi (first Il-18 services)
1959 June 15	Alma Ata–Tashkent–Baku–Rostov
1959 June 20	Moscow–Baku
1959 July 1	Moscow–Ashkhabad and Tashkent–Baku–Adler/Sochi
1959 August 2	Moscow–Frunze
1959 October 14	Moscow–Copenhagen–London (temporary Tu-104 replacement)
1959	Leningrad–Adler/Sochi and Sverdlovsk–Adler/Sochi
1960 January 5	Moscow–Bucharest–Sofia
1960 January 6	Moscow–Krasnoyarsk–Yakutsk
1960 February 20	Moscow–Krasnoyarsk (also reported as 24 November)
1960 March 2	Moscow–Dushanbe
1960 April 1	Kiev–L'vov–Bratislava–Prague
1960 April 1	Moscow–Cairo
1960 April 2	Moscow–Berlin
1960 April 10	Moscow–Kiev–Vienna
1960 April 12	Moscow–Noril'sk
1960 April 15	Moscow–Noril'sk–Tiksi (technical flight)
1960 May 25	Leningrad–Baku
1960 June 1	Moscow–Riga
1960 December 20	Moscow–Tiksi–Magadan (technical flight)

Tabso's Il-18 LZ-BES at Benina Airport, Benghazi, in April 1966 while operating Haj charters to and from Saudi Arabia. (*John Stroud*.)

1961 January 10	Moscow–Tiksi–Magadan
1961 February 3	Leningrad–Krasnoyarsk
1961 July 1	Moscow–Yerevan
1961 August 15	Moscow–Karaganda and Alma Ata–Karaganda
1961 August 17	Moscow–Anadyr'
1962 May 3	Moscow–Astrakhan
1962 May 15	Moscow–Leninabad and Moscow–Semipalatinsk
1962 June 2	Krasnoyarsk–Adler/Sochi
1962 June 29	Moscow–Khartoum
1962 September 5 or 15	Moscow–Kemerovo
1962 September 11	Moscow–Belgrade–Rabat–Conakry–Accra
1963 January 18	Moscow–Chelyabinsk
1963 February 6	Yerevan–Tashkent
1963 February 25	Moscow–Arkhangel'sk
1963 February 27	Leningrad–Arkhangel'sk and Leningrad–Murmansk
1963 April 2	Chelyabinsk–Adler/Sochi
1963 April 5	Riga–Gor'kiy–Sverdlovsk
1963 May 14	Riga–Novosibirsk
1963 May 20	Moscow–Ulan Bator
1963 May 23	Moscow–Damascus
1963 August 12	Moscow–Bamako
1963 August 15	Moscow–Gudauta
1963 December 1	Moscow–Gor'kiy and Moscow–Murmansk
1963 December 5	Moscow–Karachi
1963 December 7	Moscow–Tselinograd
1964 February 21	Moscow–Belgrade–Algiers
1964 March 14	Moscow–Tashkent–Karachi–Colombo
1964 May 15	Moscow–Yuzhno-Sakhalinsk, Leningrad–Yuzhno-Sakhalinsk, Moscow–Blagoveshchensk, Moscow–Gudauta and Moscow–Zaporozh'ye
1964 July 1	Moscow–Sukhumi
1964 July 16	Moscow–Damascus–Baghdad
1964 July 23	Moscow–Nicosia
1964 November 1	Moscow–Teheran

1964 November 15	Moscow–Volgograd
1964 November 16	Tashkent–Nukus
1964 December 1	Moscow–Nukus
1965 February 16	Moscow–Perm
1965 March 1	Frunze–Krasnoyarsk
1965 May 15	Moscow–Donetsk
1966 June 1	Arkhangel'sk–Kiev–Odessa
1966 June 28	Moscow–Beirut
1967 autumn	Moscow–Magadan nonstop (Il-18D)
	(intermediate stops not shown in every case)

ČSA

1960 January 28	Prague–Bratislava (first ČSA Il-18 service, by OK-NAA)

Deutsche Lufthansa/Interflug

1961 March 30*	Berlin–Moscow
1965 June 22*	Berlin–Cairo
1966 June 1*	Berlin–Kiev
1966 October 27	Berlin–Algiers–Bamako

Ghana Airways

1962 October 14	Accra–Kano–Cairo–Beirut (first Ghana Airways Il-18 service)
1963 February 26	Accra–Bamako–Tunis–Zürich–Warsaw–Moscow
1963 May 3	Accra–Ouagadougou–Bamako–Rabat
1963 October 30	Accra–Kano–Khartoum–Addis Ababa

Malév

1961 April 25	Budapest–London (first service by HA-MOC)

Polskie Linie Lotnicze (LOT)

1961 May 24	Warsaw–Moscow (first LOT Il-18 service, by SP-LSA)
1965 September 7	Warsaw–Frankfurt-am-Main (first service by SP-LSB)

Tabso

1966 May 5*	Sofia–Moscow

* Unconfirmed.

Ilyushin Il-20/28

The Ilyushin Il-28 was a twin-jet tactical bomber first flown in 1948. It became widely used by the Soviet Air Force and was supplied to China, Cuba, Finland, Hungary, Indonesia, Poland, Rumania and the United Arab Republic. Il-28s were also put into production in Czechoslovakia, and there were torpedo-bomber, reconnaissance and trainer versions with the designations Il-28T, Il-28R and Il-28U.

The Il-28 was not a transport aircraft, but a number of modified aircraft were used by Aeroflot to acquire jet operating experience before the introduction of the Tu-104s. It is almost certain that the examples used by Aeroflot were given the designation Il-20, but in the USSR they are still referred to as Il-28s.

These demilitarized Ilyushins were the first turbojet aircraft operated by Aeroflot and were put into service in February 1956, carrying newspaper matrices over the Moscow–Sverdlovsk–Novosibirsk route. It is not known how long this operation continued nor how many of these Ilyushins were used. The aircraft bore civil registrations, but none are known. The first flight over the route is believed to have been made by the aircraft bearing c/n 54005777.

The crew comprised pilot, radio operator and navigator, and normal operating altitude was about 12,000 m (39,370 ft) with Mach 0·74–0·75 indicated cruise.

The Il-20 (c/n 54005777) in service with Aeroflot for the carriage of newspaper matrices in 1956. This photograph is believed to show the arrival of the inaugural flight at Novosibirsk.

An Ilyushin Il-28 twin-jet bomber of the German Democratic Republic's air force. It is seen just after take-off with flaps lowered and undercarriage retracting. (*Courtesy William Green.*)

The Il-20/28 was a shoulder-wing cantilever monoplane with single fin and rudder and fully-retractable nosewheel undercarriage. The wing had an unswept leading edge and swept-forward trailing edge. Flaps were fitted between the fuselage and engine nacelles and between the nacelles and the ailerons.

The fuselage was of circular section, had a raised single-seat cockpit and typical Soviet bomber nose. There was an entrance hatch in the top of the nose on the starboard side. The military aircraft had a rear gun turret with twin 23-mm guns, and two forward-firing guns in the nose. The civil aircraft retained the forward gun ports but it is not known whether the tail turret was removed. The commercial load was carried in the bomb bay.

The fin and rudder and dihedral tailplane all had sweepback. The main undercarriage units, with single wheels, retracted forward into the engine nacelles, and the single nosewheel was backward retracting.

The engines were two Klimov VK-1 centrifugal-flow turbojets each developing 2,700 kg (5,952 lb) static thrust. Maximum fuel capacity was 7,900 litres (1,737 Imp. gal).

Span 21·45 m (70 ft 4½ in); length 17·65 m (57 ft 11 in); wing area 60·8 sq m (654·44 sq ft); track 7·4 m (24 ft 3¼ in).

Maximum fuel 6,400 kg (14,109 lb); normal take-off weight 18,400 kg (40,564 lb); maximum take-off weight 21,000 kg (46,297 lb); landing weight 14,690 kg (32,386 lb).

Speed at sea level at 18,400 kg (40,564 lb) 786 km/h (424·12 kt) (488·39 mph); maximum speed at nominal power at 18,400 kg (40,564 lb) at 5,250 m (17,224 ft) 848 km/h (457·58 kt) (526·92 mph); maximum speed at military power at 4,500 m (14,763 ft) 900 km/h (485·64 kt) (559·23 mph); take-off speed at 18,400 kg (40,564 lb) 220 km/h (118·71 kt) (136·7 mph); take-off speed at 21,000 kg (46,297 lb) 234 km/h (126·26 kt) (145·39 mph); landing speed at 14,690 kg (32,386 lb) 185 km/h (99·82 kt) (114·94 mph).

Take-off run at 18,400 kg (40,564 lb) 875 m (2,870 ft); take-off run at 21,000 kg (46,297 lb) 1,150 m (3,773 ft); initial rate of climb at 18,400 kg (40,564 lb) 15 m/sec (2,952 ft/min); landing distance at 14,690 kg (32,386 lb) 1,170 m (3,838 ft).

Ceiling 12,300 m (40,354 ft); maximum fuel range at Mach 0·9 at 10,000 m (32,808 ft) 2,180 km (1,176 n.miles).

The first prototype Il-62, SSSR-06156, at Vnukovo Airport, Moscow, probably in 1963.

Ilyushin Il-62

The Ilyushin Il-62 was first announced on 24 September, 1962, the day on which it was inspected by Prime Minister Krushchev, but it had been known for some time that the Soviet Union was building a rear-engined four-jet transport. The prototype Il-62, SSSR-06156, made its first flight in January 1963, piloted by Vladimir Kokkinaki, and it is believed that the type number, Il-62, was based on the year in which the aircraft was planned to appear.

Although the Soviet Union had built the four-engine Tu-110, the Il-62 was the first four-engine jet transport to go into production in the USSR, and the first to be operated by Aeroflot. By May 1963 the Il-62 was undergoing low-speed trials and at that time it was stated that production of a small series was about to begin. On 4 June, 1963, the prototype made its first visit to Vnukovo Airport, Moscow, and it was reported as having landed with a run of only 900–1,000 m (2,952–3,280 ft). By the summer of 1963 it was said that three prototypes were undergoing flight testing, and SSSR-06176, shown at the Paris Aero Show in 1965, was said to be the third prototype. It has, however, also been reported that there were two prototypes and three pre-production aircraft.

The prototype Il-62, SSSR-06156, on a test flight. (*Aviaexport*.)

A production Il-62, SSSR-86671, flying at the 1967 Paris Aero Show. (*R. A. Cole.*)

The Il-62 was designed for Aeroflot's intercontinental routes and its first task has been to replace some of the propeller-turbine Tu-114s. In layout the Il-62 is a sweptwing monoplane with T-tail and four rear-mounted turbofans. It closely resembles the British VC10 and Super VC10 and its dimensions are similar to those of the latter, but the Il-62 has greater wing area, take-off weight, capacity and wing sweep angle.

The wing comprises centre section and detachable outer panels. It is a three-spar structure with the torsion box formed by integrally machined skin/stringer panels. The outer two-thirds of each wing have increased chord with dog-tooth leading edge. Sweepback is 35 deg at 25 per cent chord. The ailerons are in three sections, the inner sections with spring-loaded servo tabs and the centre sections with trim tabs and servo tabs. Slotted flaps occupy the trailing edge between the ailerons and fuselage, are operated by a duplicated electrical system and have a total area of 43·48 sq m (468 sq ft). There are two hydraulically-operated spoilers situated immediately forward of the outer wing flaps. Wing leading edges are protected against ice accretion by hot air tapped from the engine compressors. Slender bullet fairings at the wing tips house the navigation lights and fuel jettison pipes. Almost the entire torsion box serves to form seven integral fuel tanks with total capacity of 100,000 litres (21,997 Imp. gal). Maximum usable fuel including that in the centre-section tank is 99,750 litres (21,942 Imp. gal).

The fuselage is an all-metal semi-monocoque structure of almost circular section. The structure is assembled from prefabricated stamped and pressed duralumin components and is designed for a 25,000 hr fatigue-free life, with 7,000–8,000 landings. Apart from the tail cone, the fuselage is pressurized to a maximum differential of 0·63 kg/sq cm (8·96 lb/sq in). The entire pressurized section is sound-proofed and thermally insulated by fibre-glass and, in some places, foam plastic.

Tail unit and starboard engines of the Il-62. The thrust reversers are seen covered. Fuselage and engine nacelle stripe and rudder are medium blue. (*John Stroud.*)

The tail unit comprises sweptback fin and rudder surmounted by a sweptback cantilever tailplane and elevators. The fin is a three-spar structure and the rudder has servo and trim tabs. The tailplane is of variable incidence and actuated by a duplicated electrical system. The elevators each have two trim tabs, one set being controlled manually by cables and the other by the autopilot. Leading edges of the fin and tailplane are protected against ice by hot air. There is a large bullet fairing at the fin/tailplane intersection and the forward part forms a surface, or suppressed, aerial.

The main undercarriage consists of two four-wheel bogies which retract inward into the wing roots. The tyres have a pressure of 9·5 kg/sq cm (135·11 lb/sq in) and all wheels are fitted with hydraulic anti-skid disc brakes which have a normal pressure of 65 kg/sq cm (924·52 lb/sq in) and an emergency pressure of 75 kg/sq cm (1,066·75 lb/sq in). The twin steerable nosewheels are forward retracting and have free fall. Tyre pressure is 7 kg/sq cm (99·56 lb/sq in). All undercarriage units and doors are hydraulically operated and have oleo-nitrogen shock absorbers. There is a hydraulically-operated twin-wheel strut in the rear fuselage and this is lowered as soon as the engines are closed down. It has been reported that a braking parachute is fitted.

The Il-62 was designed to have four Kuznetsov NK-8 bypass engines but these were not ready in time and the initial flight trials were made with 7,500 kg (16,534 lb) thrust Lyulka turbojets, and it is possible that 8,700 kg (19,180 lb) thrust Mikulin AM-3M engines were fitted at one time. However, the production aircraft have four Kuznetsov NK-8-4 turbofans each developing a take-off thrust of 10,500 kg (23,148 lb). These are two-spool engines with three-stage low-pressure compressor, eight-stage high-pressure compressor, single-stage high-pressure turbine and two-stage low-pressure turbine. Thrust reversers are fitted to the outboard engines. The engines are mounted on spectacle beams, the inboard nacelles have side structures of titanium to provide firewalls between engines and fuselage, and other titanium fire-proof bulkheads divide each nacelle into separate fire bays. The engines can be started from the TA-6 turbostarter situated in the extreme tail cone. Fuel control is automatic and each engine is fed from separate tanks by means of two booster pumps. Underwing pressure fuelling is used and there are eight overwing gravity fillers.

It is possible that in future Il-62s will be powered by the later Soloviev D-30K turbofan with 2·3:1 bypass ratio and take-off thrust of 11,500 kg (25,353 lb) in temperatures of up to 30 deg C.

The total volume of the pressurized section of the fuselage is 396 cu m (13,984 cu ft) and of the passenger cabins 163 cu m (5,756 cu ft). Fuselage length is 49 m (160 ft 9 in), width 3·75 m (12 ft $3\frac{1}{2}$ in) and height 4·1 m (13 ft $5\frac{1}{2}$ in). Cabin width is 3·45 m (11 ft $3\frac{1}{4}$ in) and headroom 2·12 m (6 ft $11\frac{1}{2}$ in). There are three underfloor pressurized holds and one aft unpressurized hold. The forward underfloor hold has a volume of 22·7 cu m (801·64 cu ft), the first of the rear underfloor holds is of 12·6 cu m (444·96 cu ft) capacity and the rear underfloor hold 6·9 cu m (243·67 cu ft). The unpressurized hold has a volume of 5·8 cu m (204·82 cu ft). Passenger entrance doors are on the port side immediately aft of the flight deck and just forward of the wing. They measure 1·4 m by 80 cm (4 ft $7\frac{1}{4}$ in by 2 ft $7\frac{1}{2}$ in) and have a sill height of 3·75 m (12 ft $3\frac{1}{2}$ in). Opposite these doors

The Il-62 SSSR-86671 flying past at the 1967 Paris Aero Show, with flaps partly lowered. (*R. A. Cole.*)

are emergency exit doors measuring 1·2 m by 46 cm (3 ft 11¼ in by 1 ft 6 in) and on each side there are two overwing emergency exits measuring 75 cm by 45 cm (2 ft 5½ in by 1 ft 5¾ in). The largest hold door is that to the forward hold and it measures 1·31 m by 1·26 m (4 ft 3½ in by 4 ft 1½ in). Cabin windows are oval, double-glazed and measure 35 cm by 24 cm (1 ft 1¾ in by 9½ in).

Five Il-62 seating configurations have been announced, with accommodation in each case being divided between fore and aft cabins. The standard economy class version has 186 seats, all six-abreast, with 72 seats in the forward cabin and 114 aft. The tourist class version has 168 seats, all six-abreast, with 66 forward and 102 aft. A first class version has 114 seats, with 45 arranged five-abreast in the forward cabin and 65 in five-abreast rows and one four-across in the rear cabin. A mixed-class layout for 85 passengers has 45 seats, five-abreast, in the forward cabin and 40 first class or de luxe seats, four-abreast, in the aft cabin. There is also a 75–100 seat business layout. Seat pitch is 750–1,020 mm (29½–47¼ in) in the normal passenger versions and 1,320 mm (52 in) in the de luxe layout. In all passenger aircraft there are two lavatories right forward on the starboard side, one amidship also on the starboard side and two aft. There is a midship galley, and forward, midship and aft wardrobes, this last able to hold 125–130 coats. Emergency inflatable chutes are stowed in underfloor compartments near the doors, and dinghies, life rafts and emergency radio packs are carried on over-water flights. Emergency oxygen is not carried

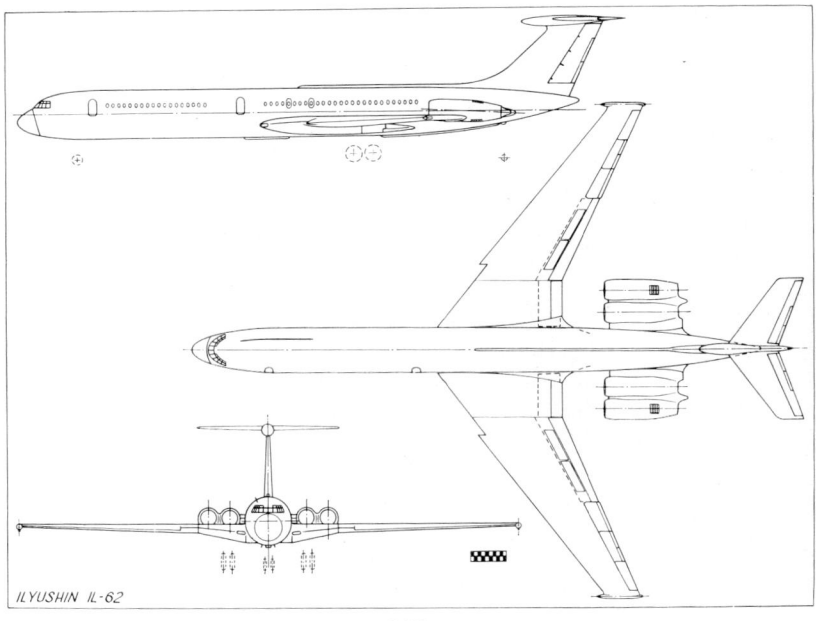

ILYUSHIN IL-62

Interior of an Il-62. (*Flight International.*)

for each occupant, but one pilot must remain on oxygen at all times at flight levels above 4,000 m (13,123 ft); in addition there is oxygen equipment for all crew members and some portable oxygen bottles carried for passenger use.

Cabin pressure is kept at sea level conditions up to a flight level of 7,000 m (22,965 ft), and at 13,000 m (42,651 ft) the interior pressure is equal to that at 2,100 m (6,889 ft). Cabin temperature is maintained at about 20 deg C. (68 deg F.).

The flight deck is laid out for two pilots, flight engineer, radio operator and navigator, and there are two additional crew seats. Navigational equipment includes VOR/ILS, Doppler and weather-warning radar. The Polyot automatic flight control system is optional and allows automatic

Part of the flight deck of the Il-62 SSSR-86671. (*Flight International.*)

approaches down to 60 m (196 ft). All flying controls are manually operated but an irreversible yaw damper is incorporated in the rudder control circuit.

The Il-62 did not enter regular passenger service with Aeroflot until 10 March, 1967. It appears to have had some problems during the test programme; there is some confirmation for this because a number of changes have been made to the wing leading edge profile.

On 2 February, 1966, an Il-62 made a proving flight over the Moscow–Khabarovsk route and on 14 February that year the first production aircraft made its initial flight. In the autumn of 1966 crew training began, and Aeroflot took delivery of its first Il-62 at the end of that year. On 1 March, 1967, Il-62s began operating cargo services between Moscow and Khabarovsk, and on 10 March they began working regular passenger and mail services over the Moscow–Khabarovsk and Moscow–Novosibirsk routes. On 14 July, 1967, they began operating Moscow–Tashkent services. On 11 July the Il-62 SSSR-86665 made a proving flight from Moscow to Montreal in 9 hr 34 min.

The Il-62 entered service on the Moscow–Montreal route on 15 September, 1967, when it began operating an extra weekly section, taking over regular weekly operation from the Tu-114 on 1 November, 1967. On 9 October they began operating between Moscow and Rome, on 14 October they replaced the weekly Moscow–Paris Tu-114 service and on 17 October they replaced the Tu-114 on the Moscow–Delhi route. In November 1967 an Il-62 flew to Washington and subsequently visited other US eastern seaboard airports in preparation for the opening of an Aeroflot Moscow–New York service. On the Moscow–Montreal service the Il-62 has a block schedule of 9 hr 50 min compared with the 12 hr 5 min schedule for the Tu-114.

The Aeroflot Il-62s are registered in the SSSR-86600 series and seven are

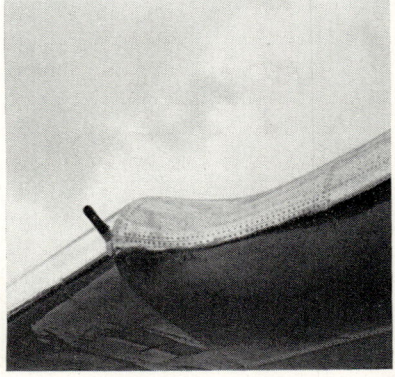

These views of the Il-62 SSSR-86671 show the dog-tooth leading edge with marked changes in profile. (*John Stroud.*)

Production Il-62 SSSR-86665 at Dorval, Montreal, in July 1967 after making a route proving flight from Moscow. (*Air Canada*.)

known, of these SSSR-86671 was exhibited and demonstrated at the 1967 Paris Aero Show in May and June.

Polskie Linie Lotnicze has stated that it expects to operate Il-62s on long-distance service after 1970, and Tabso has been negotiating the purchase of some for its projected North Atlantic services.

Direct operating costs of the Il-62 are quoted as approximately 8 US cents per ton-mile over stages of 3,000–7,000 km (1,618–3,777 n.miles) and 1·1 US cents per seat-mile over similar stages.

Span 43·3 m (142 ft 0¾ in); length 53·12 m (174 ft 3¼ in); height to top of tail 12·35 m (40 ft 6 in); wing area 282·2 sq m (3,037·57 sq ft); sweepback 35 deg at 25 per cent chord; track 6.8 m (22 ft 3½ in); wheelbase 24·5 m (80 ft 4½ in).

Equipped empty weight 67,800 kg (149,473 lb); maximum fuel 82,500 kg (181,881 lb); maximum payload 23,000 kg (50,706 lb); maximum take-off (brake release) weight 157,500 kg (347,227 lb); maximum landing weight 102,000 kg (224,871 lb).

Limiting Mach 0·9; cruising speed at 10,000—12,000 m (32,808–39,370 ft) 800–900 km/h (431·68–485·64 kt) (497·1–559·23 mph); take-off speed at maximum weight 290 km/h (156·48 kt) (180·2 mph); landing speed 220–240 km/h (118·71–129·5 kt) (136·7–149·13 mph).

Take-off run at maximum weight at sea level ISA 1,800 m (5,905 ft); take-off balanced field length at maximum weight at sea level ISA 2,950 m (9,678 ft), at sea level ISA plus 15 deg C. 3,250 m (10,662 ft); sea level rate of climb 18 m/sec (3,543 ft/min); landing run 800–1,000 m (2,624–3,280 ft); landing distance at maximum weight, with 30 deg flap and use of reverse thrust, 1,800 m (5,905 ft).

Service ceiling 13,000 m (42,651 ft).

Maximum payload range at 850 km/h (458·66 kt) (528·17 mph) with 1 hr fuel reserve, 6,700 km (3,615 n.miles); maximum fuel range at 800 km/h (431·68 kt) (497·1 mph) with 10,000 kg (22,046 lb) payload and 1 hr fuel reserve, 9,200 km (4,964 n.miles).

Kamov Ka-15

Nikolai Il'ich Kamov was one of the Soviet Union's pioneers of rotary-winged flight and his first work on helicopter design was devoted to very light helicopters for observation and communications work. His Ka-8 first flew in 1947 and was followed in 1949 by the Ka-10. Three Ka-8 prototypes were built and there were four Ka-10s followed by eight pre-production Ka-10Ms.

Kamov employed the coaxial counter-rotating type of rotor in both the Ka-8 and Ka-10 but, although this system proved satisfactory, these little helicopters were underpowered and their use proved to be severely limited.

The first Kamov helicopter to go into production was the Ka-15. Kamov had begun work on the first of his heavier helicopters in 1950 and as a result the Ka-15 began flight testing in 1952, and production was undertaken for military and naval operation. A modified version, the Ka-15M, was produced in 1953 or 1954 and this entered largescale service with Aeroflot.

The Ka-15M was a quite small helicopter with coaxial three-blade contra-rotating rotors, a cabin with two side-by-side seats, tailplane with twin fins, and a four-wheel undercarriage.

Military Ka-15, showing the coaxial contra-rotating rotors.

The fuselage was of metal construction with ply-covered forward section and stressed-skin aft. The nose and aft-sliding doors consisted mostly of transparent panels and provided good crew visibility. Immediately behind the cabin was the engine bay, aft of which the fuselage tapered quite markedly. A strut-braced narrow-chord tailplane was mounted above the rear fuselage and carried large-area end-plate twin fins and rudders. There were two main wheels, twin castoring nosewheels and a tailskid.

An agricultural Ka-15M with chemical tanks and spray-bar.

The fully-articulated rotors were of wooden construction with ply covering, and were foam-plastic filled. Rotor speed was 333 rpm.

The engine was a 255 hp Ivchenko AI-14V nine-cylinder air-cooled radial with single-speed supercharger and a cooling fan.

Flown by V. V. Vinitskiy, the Ka-15M set two closed-circuit speed records for helicopters. On 29 May, 1958, one of these helicopters achieved 162·784 km/h (101·148 mph) over a 100 km (62 mile) circuit, and on 6 May, 1959, covered 500 km (310 miles) at 170·445 km/h (105·91 mph).

The Ka-15M was produced in a number of versions and used by Aeroflot for crop-spraying and general agricultural work, for patrolling powerlines and gas pipelines, for mail carriage and as an ambulance. The Ka-15M is known to have formed part of the equipment of Aeroflot's Moldavian and Ukrainian Directorates. In spite of the large numbers which have been used by Aeroflot, many of which are thought still to be in operation, only SSSR-L0365 is known. Some of these helicopters have been supplied to collective farms.

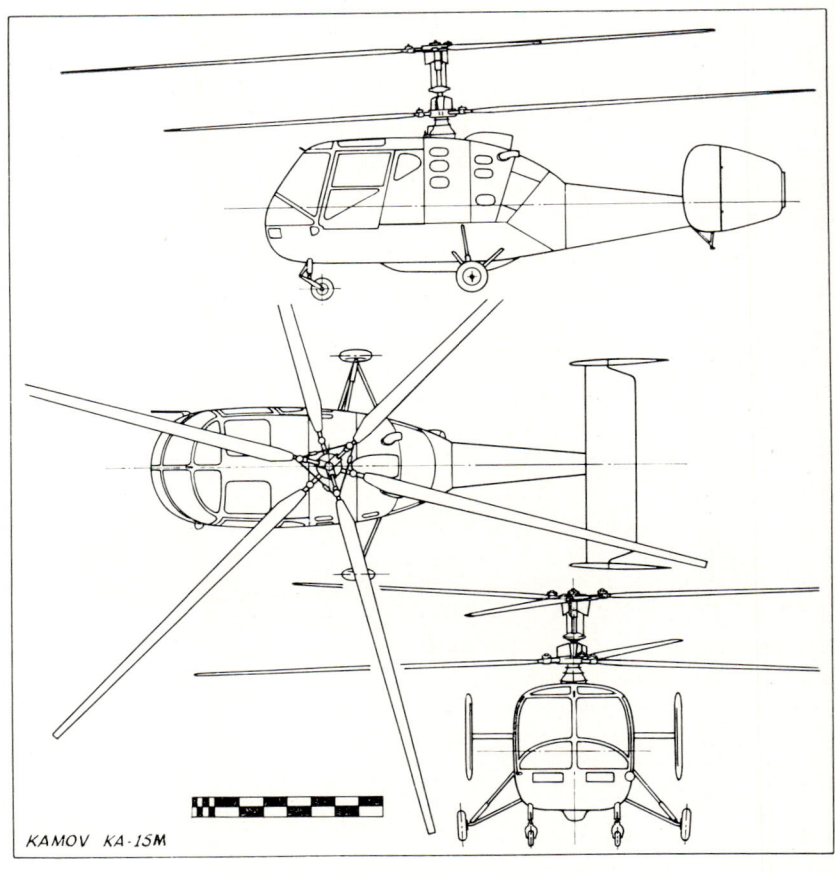

Ka-15M

Rotor diameter 9·96 m (32 ft 8 in); length 6·23 m (20 ft 5¼ in); height to top of rotor head 3·35 m (11 ft). Empty weight 990 kg (2,182 lb); fuel 130 kg (286 lb); payload 250 kg (551 lb); take-off weight 1,410 kg (3,108 lb). Maximum speed 150 km/h (80·94 kt) (93·21 mph); cruising speed 125 km/h (67·45 kt) (77·67 mph); service ceiling 3,000 m (9,842 ft); hover ceiling 680 m (2,230 ft); maximum range 390 km (210 n.miles).

Band width for spraying 10–20 m (32–65 ft), for dusting 30 m (98 ft); normal agricultural operating height 5 m (16 ft); agricultural work capacity 50 hectares/hr (123·55 acres/hr).

Kamov Ka-18

The Kamov Ka-18 helicopter is a direct development of the Ka-15 and employs the same rotor and transmission as the earlier type. The first Ka-18 was completed in 1956 and flight trials began in the following year.

The main difference between the Ka-15 and Ka-18 is in the fuselage, which in the latter has been lengthened and can carry pilot and two or three passengers. The fuselage is a welded steel-tube structure with light-alloy-skinned front section and with a monocoque tail boom. The tail unit resembles that of the Ka-15 but the fins and rudders are of modified shape and greater chord.

The original engine was the 255 hp Ivchenko AI-14V, but in 1960 this was replaced by the 275/280 hp AI-14VF which increased the payload by up to 100 kg (220 lb) and the ceiling by 300–500 m (984–1,640 ft).

Ka-18s are in service with Aeroflot on a wide range of duties and there are three basic variants. One version is used for mail carriage or the carriage of two or three passengers and baggage; another is an ambulance capable of carrying pilot, one stretcher case and a medical attendant; and the third variant is the agricultural type. In addition to these duties, Aeroflot uses Ka-18s for forestry work, mineral exploration and ship escort.

The Ka-18 is equipped for night and bad weather operation, has deicing, can be fitted with inflatable pontoons and have auxiliary fuel tanks.

Aeroflot Ka-18 SSSR-L0005.

These helicopters are used in quantity by Aeroflot, and the Moldavian and Ukrainian Directorates are known to be among those equipped with this type. A prototype Ka-18 was registered SSSR-L0005 and others known are SSSR-06137, 30204 (an agricultural aircraft), 64570, 64572 and 86606.

A feature of the Ka-18 is its easily removable rotor blades so that it can be more easily housed, particularly on board ship when used for ice reconnaissance. The blades can be removed in 5 min by two people.

Rotor diameter 9·96 m (32 ft 8 in); fuselage length 7·03 m (23 ft 0¾ in); height to top of rotor head 3·34 m (10 ft 11½ in); track of main wheels 2·84 m (9 ft 3¾ in). Empty weight 1,060 kg (2,436 lb); fuel 97 kg (213 lb); payload 240 kg (529 lb); take-off weight 1,480 kg (3,262 lb). Maximum speed 150 km/h (80·94 kt) (93·21 mph); cruising speed at 490 m (1,607 ft) 110–120 km/h (59·36–64·75 kt) (68·35–74·56 mph); service ceiling 3,250 m (10,662 ft); range 165 km (89 n.miles).

Kamov Ka-25K

A new Kamov helicopter made its first appearance at the 1967 Paris Aero Show. This was the Ka-25K utility crane which is a development of the Ka-20 anti-submarine helicopter which was itself first seen in public at the Soviet Aviation Day display in 1961. The Ka-25K, SSSR-21110 (c/n 070601-K), shown at Paris bore the date 30 December, 1966, on its rotor blades and was said to be a new type then still in the test stage.

The blue and grey Ka-25K SSSR-21110 at the 1967 Paris Aero Show. The rotor-folding power-lead can be seen plugged in to the fuselage side. Beneath the nose is a pilot's cabin from which the helicopter is controlled when working as a crane.

The Kamov Ka-25K flying at Le Bourget in June 1967. (*J. M. G. Gradidge.*)

The Ka-25K is much larger than the Ka-15 and Ka-18, is powered by two 900 ehp Glushenkov turboshaft engines and has power-folding of its coaxial contra-rotating three-blade rotors.

The fuselage is quite deep and has a cabin measuring 3·95 m (12 ft $11\frac{1}{2}$ in) in length, 1·25 m (4 ft $1\frac{1}{4}$ in) in height and about 1·4 m (4 ft $7\frac{1}{4}$ in) in width. Removable fold-up seats for eight passengers are provided along the starboard wall, and there are a further four fold-up seats on the port side forward of the aft-sliding main door which measures 1·2 m by 1·1 m (3 ft $11\frac{1}{4}$ in by 3 ft $7\frac{1}{4}$ in). There are two large windows in each side of the

The rotor head and folded blades of the Ka-25K SSSR-21110. The date 30 December, 1966, is painted on the blade roots. (*John Stroud.*)

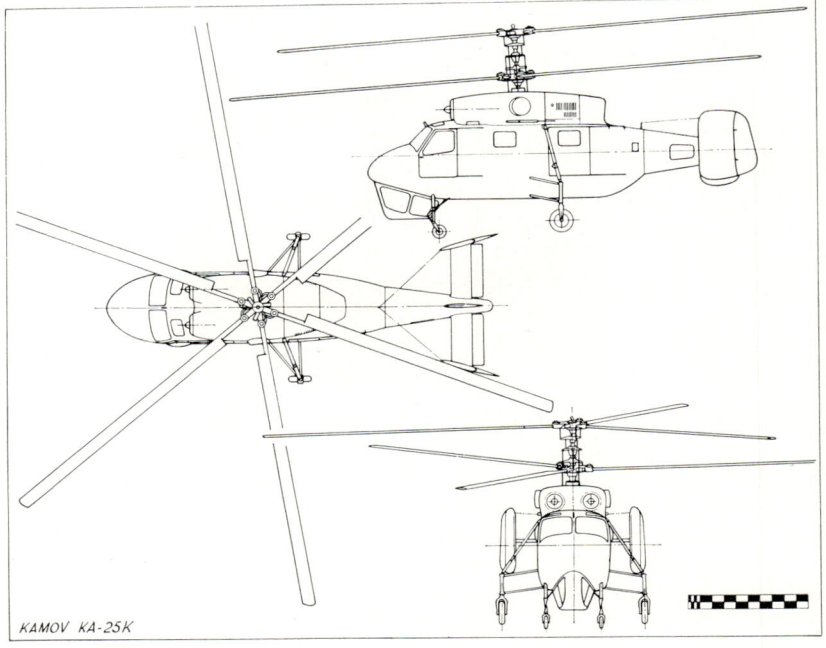

KAMOV KA-25K

cabin, the aft one on the port side being in the door. There is a hatch in the floor and above this can be placed a hydraulic hoist for carrying suspended cargo loads. The pilots' cabin, with its own aft-sliding doors, has side-by-side seats and can be fully equipped for day and night and bad weather operation. Beneath the nose is a small glazed cabin with backward-facing seat, from which the Ka-25K can be flown while operating as a crane. There is a landing light in the nose of the under-cabin, and the main flight deck windows have windscreen wipers.

The undercarriage comprises two main wheels and twin castoring nose-wheels. The tail consists of a narrow-chord tailplane with even narrower elevators, a central fin both above and below the fuselage and twin fins and rudders which are sharply toed-in towards their leading edges.

The twin-turbines are mounted above the cabin and have side efflux outlets. Working platforms are built into the base of the engine cowlings immediately beneath the rotor heads. Small protuberances near the tip of each rotor blade may be ice detectors.

Power for rotor-blade folding appears to be supplied via a cable from a plug-in position on the fuselage low down on the port side.

The Ka-25K has an all-metal structure. The blue and grey example shown in Paris bore Aeroflot's name on the fuselage and was most likely the first prototype.

Rotor diameter 15·74 m (51 ft 7¾ in); overall length with rotors running 15·74 m (51 ft 7¾ in); fuselage length 9·83 m (32 ft 3 in); height to top of rotor head 5·37 m (17 ft 7¼ in); width with rotors folded 3·76 m (12 ft 4 in); track 3·52 m (11 ft 6½ in) main wheels, and 1·41 m (4 ft 7½ in) nosewheels.

Empty weight 4,200 kg (9,259 lb); useful load including pilot and fuel 2,900 kg (6,393 lb); maximum payload 2,000 kg (4,409 lb); normal take-off weight in ISA 7,100 kg (15,652 lb); maximum take-off weight 7,300 kg (16,093 lb).

Maximum speed 220 km/h (118·71 kt) (136·7 mph); cruising speed 195 km/h (105·22 kt) (121·16 mph); service ceiling 3,500 m (11,482 ft); hover ceiling at 7,100 kg (15,652 lb) about 750 m (2,460 ft); range with 1,750 kg (3,858 lb) payload 400 km (215 n.miles); maximum range with 1,000 kg (2,204 lb) payload and fuel reserve 650 km (350 n.miles). The Ka-25K can carry a 2,000 kg (4,409 lb) slung load over a stage of 50 km (26·9 n.miles) and return to base without refuelling when operating at a take-off weight of 7,100 kg (15,652 lb).

Kamov Ka-26

The Kamov Ka-26 was first announced in January 1964 and began flight trials some time in 1965. It was first seen in public in September 1965 when it was put on show at an agricultural exhibition in Sokolniki Park, Moscow.

N. I. Kamov and V. I. Sorin were mainly responsible for the design of the Ka-26 which was produced to meet the requirements of the Soviet State Scientific Research Institute, and during work on the project Kamov's organization collaborated closely with Aeroflot.

A prototype Ka-26 undergoing tests as an agricultural aircraft at Krasnodar in the Kuban in 1966.

The blue and white Ka-26 SSSR-26002 flying with passenger and cargo pod attached.

Like the Ka-15, Ka-18 and Ka-25, the Ka-26 employs the coaxial contra-rotating rotor system but in most respects is completely different. The extensive experience of coaxial rotor systems gained with the Ka-15 and Ka-18 was fully used in design of the new helicopter, which was required to carry a chemical load of 600–800 kg (1,322–1,763 lb) and be capable of spraying one hectare (2·471 acres) of crops for a cost of not more than 3 roubles.

What Kamov produced in fact was a most ingenious multi-purpose helicopter capable of almost any task except feeding itself.

The Ka-26 consists of a basic airframe to which can be attached the means of carrying out various duties. The basic structure consists of two piston engines, transmission and coaxial rotor system, four-wheel undercarriage, twin fin tail unit supported on twin fuselage booms, and control cabin. The commercial load is carried immediately behind the control cabin and on the centre of gravity.

Among the duties of which the Ka-26 is capable are: passenger, mail and cargo transport; agricultural work including chemical spraying and dusting; mapping, geodesic and geological survey; forest fire patrol; ice reconnaissance; gas and oil pipeline and electric transmission line patrol; fish spotting; rescue; erection and building work; carriage of firefighting teams and geological and scientific expeditions.

To accomplish these varied tasks a number of different load carriers can be fitted. For passenger and cargo transport a cabin-pod can be attached. This is a simple structure having a volume of 3·2 cu m (113 cu ft), floor-level length of 1·83 m (6 ft) and width of 1·25 m (4 ft 1¼ in). Its interior height is 1·4 m (4 ft 7¼ in). This unit has slightly bulged sides which, as well as the roof, are strengthened by external stiffeners. There are two large windows in each side. The rear of the cabin consists of two full-height outward-opening doors each incorporating a very large window. In the forward end of the cabin a doorway provides access to the control cabin. The floor, which incorporates a jump hatch also serving as an emergency exit, is a sandwich with outer panels of fibre-glass and a core of aluminium sheets. The walls are of synthetic fire-resistant material which provides thermal and noise insulation. Fold-up bench-type seats along the side walls provide accommodation for a total of six passengers. Padded areas beneath the windows serve as back rests, and seat-belts are provided. With seats folded against the walls there is space for 700 kg (1,543 lb) of cargo. This cabin-pod is attached to the Ka-26's airframe by six quick-disconnect locks, and can be easily fitted or removed by a team of three.

For agricultural work there are two configurations. When used for spraying, the Ka-26 is fitted with a 900 kg (1,984 lb) capacity fibre-glass hopper and a wide-span spray-bar with adjustable atomizer-jets, and on SSSR-21111 shown at the 1967 Paris Aero Show there was an additional spray-bar behind the tail unit. For dusting, the same hopper is used but attached to it are two discharge nozzles. Discharge is at the rate of 14 litres (3·07 Imp. gal) per second under a pressure of 5 atmospheres. Eddies caused by the coaxial rotor system are said to cause uniform spreading of chemicals on upper and lower surfaces of the leaves of the plants being treated.

The cargo duties of the Ka-26 can be undertaken with a cargo platform fitted with lashing points or a remote-controlled winch with sling and hook. In either configuration a payload of 900 kg (1,984 lb) can be carried.

The grey, red and white Ka-26 SSSR-21111 at the 1967 Paris Aero Show. It is seen with hopper and spray-bar and an additional spray-bar aft of the tail unit. (*John Stroud.*)

For ambulance work the Ka-26 can carry two stretcher cases, two sitting cases and a medical attendant. A winch, capable of lifting 150 kg (330 lb), is fitted for rescue work.

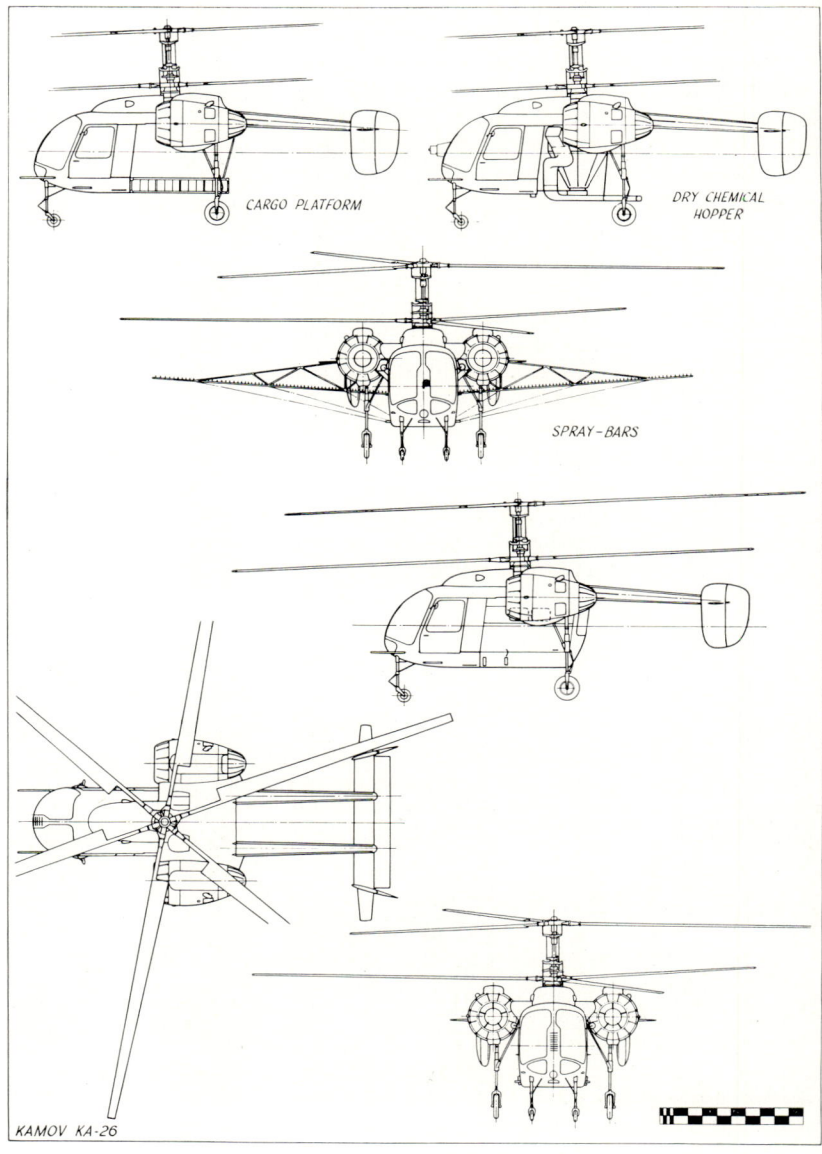

The control cabin has two side-by-side seats and one of these can be used as a passenger seat, but dual controls can be fitted in an hour and a half.

In the construction of the Ka-26 widescale use has been made of plastic and synthetic glues. The rotor blades, six in all, have been made of fibreglass, and high standards of manufacturing precision ensure complete interchangeability of blades.

The Ka-26 as an agricultural aircraft with hopper and distributors for dry chemicals. (*Aviaexport*.)

One of the biggest problems encountered with any aircraft used for agricultural work is the toxic nature of the loads carried and the corrosive effect of these loads. Glue-welded and glue-riveted seams have been widely used to produce an airtight structure, thus keeping out of the structure any harmful chemicals or dust. The use of plastics, which represent 20 per cent of the structure weight, have not only made for a longer aircraft life, they have cut structure weight.

A blower and air-filter unit is used to protect the crew by ensuring that the air is clean and that there is a slight pressure differential. This unit is apparently the protuberance on the nose of the agricultural versions of the Ka-26.

The control cabin can be heated and lighted. A very neat instrument panel is centrally located and if required the Ka-26 can be fitted with autopilot, radio compass, radio altimeter, short-wave radio and a navigational system for night and bad weather operation. Rotor blades and control cabin windows have ice protection, an automatic radio-isotope ice-warning system being employed.

The basic structure of the Ka-26 consists of a short-span centre section to which the engines are attached. The forward part of the centre section carries the control cabin, and running aft from the centre section are two tubular tail booms which carry the tailplane. The centre section also carries the main undercarriage units and houses the fuel.

Twin fins and rudders are fitted and the control runs for the rudders are carried inside the port tail boom. It is not known for certain whether there is a limited-travel elevator but there appears to be. A tail modification was the extension of the tailplane outboard of the vertical fins. The fins and rudders are sharply toed-in towards their leading edges.

The Ka-26 fitted with spray-bar, and, alongside, the passenger pod and a loaded cargo platform. (*Aviaexport*.)

The engines originally fitted to the first prototype were 275 hp Ivchenko AI-14VF nine-cylinder air-cooled radials, but these were later reported to have been replaced by the 325 hp AI-14FR. Production aircraft are powered by two 325 hp M-14V-26 engines said to have been designed by I. M. Vedeneev; it would appear in fact that these are AI-14FRs modified specially for the Ka-26 by Vedeneev. The Ka-26 has single-engine flight ability.

The Ka-26 can spray 30 hectares (70 acres) an hour, and its overhaul life is stated to be 1,000 hr.

One of the great disadvantages of using helicopters for agricultural work is the short duration for which they are needed. Spraying and dusting have to be done at specific times and the small number of flying hours involved makes the helicopter an uneconomic vehicle for this work. It is because of this limited period of use on such duties that the Ka-26 has been designed to have such a wide range of work possibilities, and its ability to undertake such a variety of tasks could make it a most successful venture.

So far five Ka-26s are known: the brown and white SSSR-26001 which was probably the first prototype, the blue and white SSSR-26002, SSSR-26003, SSSR-26004, and the red, white and grey SSSR-21111.

Rotor diameter 13 m (42 ft 7¾ in); fuselage length 7·75 m (25 ft 5 in); width excluding rotors 3·64 m (11 ft 11½ in); height to top of rotor head 4·05 m (13 ft 3½ in); distance between upper and lower rotor 1·17 m (3 ft 10 in); track 2·42 m (7 ft 11¼ in) main undercarriage, and 90 cm (2 ft 11½ in) nosewheels; wheelbase 3·48 m (11 ft 5 in).
Weights:
Stripped helicopter. Operating empty weight 1,950 kg (4,299 lb).
Passenger–cargo version. Operating empty weight 2,120 kg (4,673 lb), fuel 285 kg (628 lb), payload 595 kg (1,311 lb), maximum take-off weight 3,000 kg (6,613 lb).
Agricultural spraying version. Operating empty weight 2,150 kg (4,740 lb), fuel 100 kg (220 lb), payload 800 kg (1,763 lb), maximum take-off weight 3,050 kg (6,724 lb).
Agricultural dusting version. Operating empty weight 2,160 kg (4,762 lb), fuel 100 kg (220 lb), payload 900 kg (1,984 lb), maximum take-off weight 3,160 kg (6,966 lb).
Cargo version with platform. Operating empty weight 2,000 kg (4,409 lb), fuel 100 kg (220 lb), payload 900 kg (1,984 lb), maximum take-off weight 3,000 kg (6,613 lb).
Cargo version with hook. Operating empty weight 2,050 kg (4,519 lb), fuel 100 kg (220 lb), payload 900 kg (1,984 lb), maximum take-off weight 3,050 kg (6,724 lb).
Maximum speed 170 km/h (91·73 kt) (105·63 mph); maximum cruising speed 130 km/h (70·15 kt) (80·78 mph); economic cruising speed 100 km/h (53·96 kt) (62·13 mph); service ceiling up to 3,000 m (9,842 ft); hover ceiling outside ground effect 800 m (2,624 ft) at 3,000 kg (6,613 lb), 1,800 m (5,905 ft) at 2,600 kg (5,732 lb); hover ceiling in ground effect 800 m (2,624 ft) at 3,160 kg (6,966 lb), 1,800 m (5,905 ft) at 2,800 kg (6,172 lb); range at 500 m (1,640 ft), with 7 passengers and 30 min fuel reserve, 400 km (215 n.miles); endurance at economic cruising speed 3½ hr; maximum range, with additional tankage, up to 1,200 km (647 n.miles); fuel consumption at 100 km/h (53·96 kt) (62·13 mph) 40 kg/hr (88 lb/hr).

Pre-production Mi-1s, probably photographed at Tushino in 1951.

Mil Mi-1

At the present time there are only two Soviet aircraft design bureaux responsible for helicopter work, one under N. I. Kamov and the other under Mikhail Mil. Kamov concentrates on designs employing coaxial contra-rotating rotors but the Mil designs are all classical helicopters with single main rotor and tail anti-torque rotor.

Mikhail Mil had his first contact with helicopter design work as far back as 1931, but this was interrupted and not resumed until at the end of the 1939–45 war he became the head of TsAGI's Scientific Research Laboratory for Rotorcraft. In 1947 an official specification was issued for a three-seat communications helicopter, and Mil set up his own design bureau in order to compete with others in meeting this specification.

Mil's design was the GM-1 on which work began late in 1947. Three prototypes were built, with the first being ready for trials in September 1948. It was test flown by M. K. Baikalov but on an early flight went out of control and crashed. However, the second and third prototypes completed their factory and State trials and in September 1949 series production was recommended. The production helicopter bore the designation Mi-1 and the type was the first Soviet helicopter to be produced in quantity. Pre-production examples were seen at the Soviet Aviation Day display at Tushino in 1951 and Mi-1s entered service with the Soviet Air Force before the end of the year.

In layout the Mi-1 was a conventional single-rotor helicopter with deep forward fuselage, slim tail boom, three-blade main and tail rotors and non-retractable nosewheel undercarriage. It had two small controllable-incidence horizontal fins near the rear of the tail boom and a long tailskid to protect the rear rotor.

The main fuselage was a metal-skinned steel-tube structure and the tail boom was a semi-monocoque metal structure. The rotor blades were tapered and of composite construction, pitch-change was achieved by tilting the main rotor, and both main and tail rotors had a deicing system. The engine was a 575 hp Ivchenko AI-26V or AI-26GRF seven-cylinder radial with fan cooling. The main wheels were fitted with brakes and the nosewheel was steerable. Standard internal fuel capacity was 240 litres (52 Imp. gal). There was accommodation for pilot and two or three passengers.

A programme of modifications and improvements resulted in a steady extension of overhaul life from the original 150–200 hr to 300 hr in 1952, 500–600 hr in 1956–57 and 1,000 hr in 1959–60. It was reported that by 1966 rotor blade life was 2,000 hr and rotor head 1,000 hr.

The Mi-1 remained in production for 12 years, was built in large numbers and appeared in several versions. The first variant was the Mi-1U trainer with dual control and then in 1954 came an ambulance variant with removable side panniers. This ambulance version had increased weight and was fitted with a four-blade rotor. Its trials were satisfactory but it was not put into production although other versions did serve as ambulances.

An agricultural version was the Mi-1NKh (Narodnoye Khozyaistvo–national economy) all-weather multi-purpose model for agricultural work, forestry patrol, ambulance work and mail carriage. As an agricultural

A Dosaaf Mi-1 in dark green and grey-blue paint scheme with red star marking.

aircraft the Mi-1NKh could carry two 250 litre (55 Imp. gal) external tanks, a spray-bar and dusting nozzles. Other production Mi-1s were the Mi-1A and Mi-1T which are believed to have differed from standard Mi-1s mainly in instrumentation and equipment.

In 1958 Mi-1s were fitted with various types of flotation gear so that they could work with the whaling fleets. One of these helicopters, fitted with four cylindrical bag floats, made 184 whale-spotting flights and flew 42,250 km (26,253 miles) in 338 hr between 22 October, 1958, and 16 May, 1959. Mi-1s also established numerous FAI records. These included distance in a straight line of 794·918 km (493·938 miles) in 5 hr 22 min on 19 March, 1958, with F. I. Belushkin as pilot, and on 12 March, 1959, the same pilot

An Aeroflot Mi-1, SSSR-20280, with auxiliary fuel tanks, taking-off from Murmansk. On the ground are An-2s and Yak-12s.

set an altitude record of 6,700 m (21,981 ft). V. V. Vinitskiy set a 100 km (62 mile) closed-circuit record of 210·535 km/h (124·6 mph) on 21 May, 1959, having achieved 196·452 km/h (122·069 mph) over a 500 km (310 mile) course on 19 May that year.

The last known version of the Mi-1 family was the Mi-1 Moskvich of 1961. This had all-metal rotor blades, improved cabin, and hydraulic control system. It is not known whether it was put into production.

In Poland, the Mi-1 was put into production under the designation SM-1 and the first example flew in the spring of 1956. An improved version, the SM-1W, appeared in 1960 and was built for domestic use and export in several variants. There was the SM-1W basic three-passenger type, the SM-1WS ambulance, the SM-1WZ agricultural version and the SM-1WSZ trainer. In 1960 a Polish-designed development was flown, this was the SM-2 with generally improved forward fuselage. When Polish production of the Mi-1 began, it was undertaken by Wytwornia Sprzetu Komunikacyjnego at Swidnik; but in 1957 W.S.K. Swidnik was

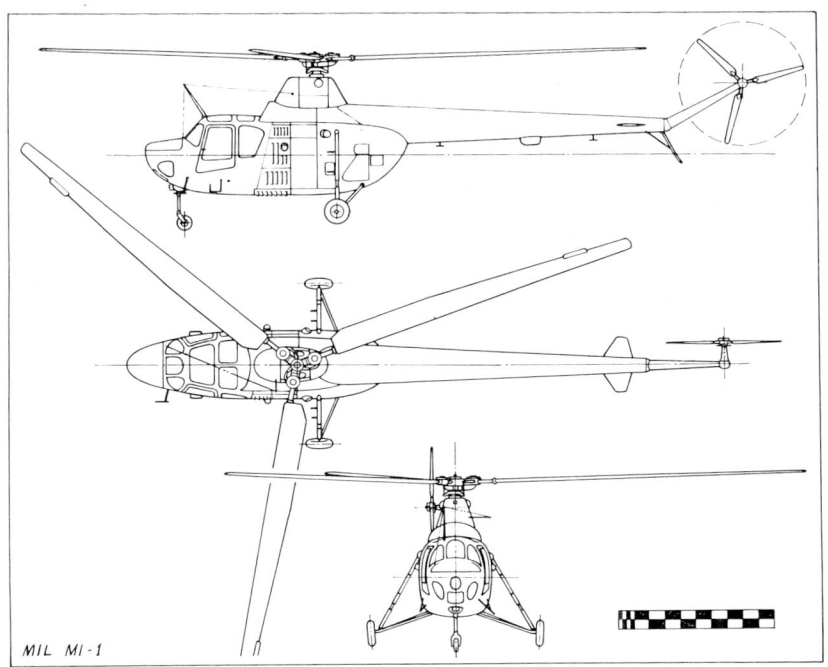

MIL MI-1

renamed Polskie Zakłady Lotnicze, thus reviving the well-known prewar initials PZL.

The Soviet Air Force employed large numbers of Mi-1s, mostly the

The Mi-1 Moskvich SSSR-68131 at Vnukovo Airport, Moscow. This helicopter is believed to have been painted white with red trim and blue lettering.

Mi-1NKh version but also the Mi-1T, and in May 1954 the Mi-1 entered service with Aeroflot. With Aeroflot, Mi-1s were used for agricultural work, forest patrol, geological exploration, whale-spotting, local service operation for carriage of passengers and cargo, and for ambulance work. The Kirgiz Directorate used Mi-1s in areas where mountain ranges made road building a major problem. The Moldavian and Ukrainian Directorates used Mi-1s, and it is likely that they were operated by most sections of Aeroflot and equally likely that numbers remain in service.

ČSA's Agrolet division had at least one Mi-1, Tabso used the type and Yemen Airlines is known to have had two. Mi-1s were supplied to several air forces including that of Finland.

Main rotor diameter 14·5 m (47 ft 7 in); disc area 160·5 sq m (1,727·61 sq ft); length with rotors running 17 m (55 ft 9¼ in); fuselage length 12·11 m (39 ft 8¾ in); tail rotor diameter 2·5 m (8 ft 2½ in); height 3·3 m (10 ft 10 in); track 3·3 m (10 ft 10 in); wheelbase 3·17 m (10 ft 5 in); fuselage ground clearance 40 cm (1 ft 3¾ in).

Empty weight 1,900 kg (4,188 lb); fuel 175 kg (385 lb); maximum payload 650 kg (1,433 lb); maximum chemical load 300 kg (661 lb); normal take-off weight 2,500 kg (5,511 lb); maximum take-off weight 2,550 kg (5,621 lb).

Maximum speed 170 km/h (91·73 kt) (105·63 mph); cruising speed 135 km/h (72·84 kt) (83·81 mph); long-range cruising speed 90 km/h (48·56 kt) (55·92 mph); maximum vertical rate of climb at sea level 6·5 m/sec (1,279 ft/min); maximum vertical rate of climb at sea level at rated power 4 m/sec (787 ft/min); maximum rate of climb at 3,000 m (9,842 ft) at rated power 3·5 m/sec (688 ft/min); economic climbing speed 90 km/h (48·56 kt) (55·92 mph); service ceiling 4,000 m (13,123 ft); hover ceiling 3,000 m (9,842 ft); normal range 350 km (188 n.miles); maximum range 590 km (318 n.miles).

Weights and performance of different versions varied, and the Mi-1A had a smaller rotor with a diameter of 14·34 m (47 ft 0½ in).

Mil Mi-2 (V-2)

Although very large numbers of helicopters have been built with piston engines supplying the power, it was the advent of the gas-turbine which opened up the future for the helicopter as a reasonably economic vehicle, and the Soviet helicopter designer Mikhail Mil was quick to take advantage of the potential of the shaft-turbine. Mil's first turbine-powered helicopter was his very large Mi-6 but he also employed turbine engines to transform the small helicopter into a much more useful vehicle.

In the piston-engine powered Mi-1 the engine and transmission accounted for 25 per cent of the total structure weight and the main rotor for 12 per cent. Increases in lift for the piston-engined helicopter always

The Mi-2 SSSR-06180 with auxiliary fuel tanks. This is believed to have been one of the prototypes.

brought about even greater increases in weight; the turbine engine, on the other hand, provides much greater power and reduced weight.

The availability of the turbine engine led Mil to design a much improved version of the Mi-1. The new helicopter, known as the Mi-2 and also as the V-2 (Vertolet—helicopter), was powered by two Izotov shaft-turbines each developing 400 shp. These two engines weigh less than half the weight of the single 575 hp Ivchenko AI-26 piston engine used to power the Mi-1.

The shaft-turbines also had another enormous advantage. They could be placed above the cabin thus leaving much more room for commercial load while at the same time greatly easing c.g. problems. By placing the engines well forward it was possible to position the payload below the axis of the main rotor and thus vary the load carried without encountering balance problems. Centre of gravity range about the rotor axis is 185 mm (7·28 in) forward and 10 mm (0·39 in) aft.

The Mi-2 SSSR-06180 with modified undercarriage. It was used in this form to set a closed-circuit speed record on 20 June, 1965.

The change from piston engine to turbine power completely changed the economics of the Mi-1 category helicopter for it enabled the payload to be more than doubled while keeping to approximately the same overall dimensions.

In layout the Mi-2 is an orthodox single-rotor helicopter with tail anti-torque rotor and non-retractable nosewheel undercarriage. The fuselage is a stressed-skin metal-bonded structure with deep forward section and tubular tail boom on which are mounted variable-incidence horizontal stabilizers. The main rotor has three parallel-chord blades which comprise

The Polish-built Mi-2 SP-PSC at the 1967 Paris Aero Show. (*John Stroud.*)

a pressed load-carrying spar to which are attached 20 narrow aerofoil sections which appear to consist of an aluminium honeycomb. No rivets or bolts are used in the assembly of the blades, all components being joined by a metal adhesive known as VK-3. The joints have a strength of 250 kg/sq cm (3,555 lb/sq in) and the blades a life of 2,000 hr. The rotor blades have electro-thermal leading edge deicing. A warning system is provided to detect any fatigue cracking in the spars. Hydraulic blade oscillation dampers are used in place of the friction dampers on the Mi-1. Although Mil has stressed the fact that no tail rotor failures have occurred in the Mi-1s, it was decided to construct the two-blade tail rotor of the Mi-2 from metal, with a light alloy spar, metal skin and honeycomb core—all metal bonded.

The Mi-2's undercarriage consists of main units with oleo-pneumatic shock absorbers, and twin nosewheels. There is a tall tailskid. Pneumatic brakes on the main wheels make take-offs possible from slopes of up to 30 deg.

The engines in the Mi-2 are described as Izotov GTD-350 free-turbines with 500 hr overhaul life. In fact GTD are the initial letters of the Russian words for gas, turbine and engine, and therefore the quoted designation may simply be a general description. This type of engine has a compressor with seven axial stages and one centrifugal stage. There is a single-stage compressor turbine with air-cooled disc and a two-stage constant-speed power turbine. The engine, which is also in production in Poland, measures 1·35 m (4 ft 5 in) in length, 520 mm (20·47 in) in width and 630 mm (24·8 in) in height. Its dry weight excluding jet pipe and accessories is 135 kg (297 lb). Take-off rating is 400 hp, nominal rating 320 hp and cruise ratings are 235–285 hp. Cruise consumption is 98–128 kg/hr (216–281 lb/hr).

The engines are mounted side-by-side above the front of the cabin and have direct air intakes and above them is an air intake for the cooling fan. The transmission drives the main rotor, tail rotor, cooling fan and the auxiliary systems mounted on the main reduction gearbox. The entire installation is extremely neat. There is a main internal fuel tank of 600 litres (131 Imp. gal) capacity, and two 238 litre (52 Imp. gal) external cylindrical tanks can be attached, one to each side of the fuselage, to bring total capacity to 1,076 litres (235 Imp. gal).

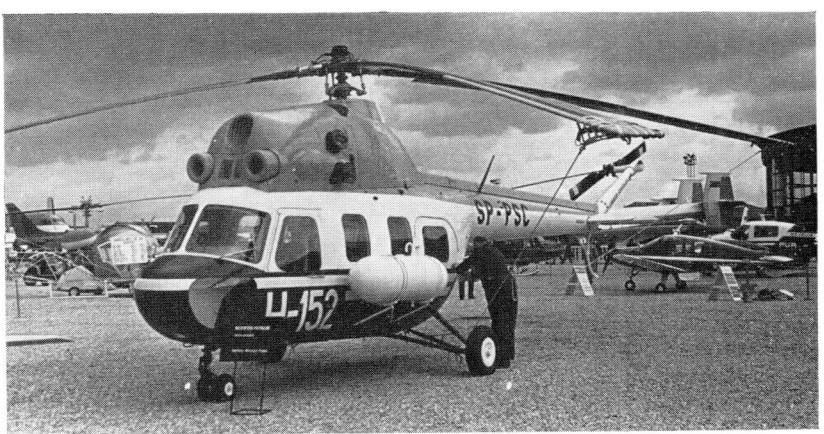

The Mi-2 SP-PSC with auxiliary fuel tank. (*John Stroud.*)

The Mi-2 is produced in passenger, cargo, agricultural and ambulance versions. Total cabin length is 4·47 m (14 ft 8 in), floor width is 1·2 m (3 ft 11¼ in) and height is 1·4 m (4 ft 7¼ in). In the passenger version there is a central bank of back-to-back seats for three forward-facing and three aft-facing passengers, there is a single forward-facing seat on the starboard side against the rear bulkhead and an eighth passenger can be carried alongside the pilot. The seats can be easily removed to provide space for 700 kg (1,543 lb) of cargo. There is a large door aft on the port side and another forward on the starboard side. A slung cargo load of up to 800 kg (1,763 lb) can be carried by an under-fuselage hook, and an electric hoist of 120 kg (264 lb) capacity can be fitted on the port side aft of the cabin. As an ambulance the Mi-2 can carry four stretcher cases and a medical attendant.

For agricultural work the entire load is carried externally. A 500 litre (109 Imp. gal) tank is carried on each side of the fuselage and can be used for dry chemical or liquid. Funnel-type distributors are attached for dusting, and for dispensing liquid chemicals a wide-span spray-bar is fitted. Maximum chemical load is 1,000 kg (2,204 lb).

A Polish-built Mi-2 with hoppers and distributors for dry chemical dusting.

The Mi-2 was first announced in the autumn of 1961 and it is reported to have made its first flight during that year. It was demonstrated to members of the Soviet Government on 25 September, 1962, and production began in 1963. In 1966 all civil production was transferred to Poland where it is built by Polskie Zakłady Lotnicze (PZL).

On 14 May, 1963, an Mi-2 set a 100 km (62 mile) closed-circuit record with a speed of 253·818 km/h (157·7 mph) but on 20 June, 1965, Tatiana Roussian flew the specially modified SSSR-06180 over a 100 km (62 mile) circuit at 269·38 km/h (167·381 mph). On that occasion the Mi-2 was fitted with spat-enclosed main wheels of very small diameter and a nose skid in place of the normal twin nosewheels.

In 1964–65 experimental models of the Mi-2 were used for fertilizing winter crops in the Moscow area and as a result the grain harvest was increased by 200 kg per hectare (about 176 lb an acre). In the spring of 1966 an experiment was made in the Kaluga district during which four Mi-2s fertilized up to 250 hectares (617 acres) a day in 5–6 hr, with a spread of 150 kg per hectare (about 132 lb per acre). So that productivity could be compared, An-2 biplanes took part in the same experiment, and it was found that the transport cost of fertilizers was considerably reduced by the use of helicopters because they could be loaded at the warehouse. It has been found that in some parts of the USSR there is year-round work for agricultural helicopters; but where full-time provision of helicopters would not be worthwhile the Mi-2, unlike the Mi-1, will be able to undertake profitable transport work as an alternative. With the Mi-1 it was found that the payload was too small to achieve economic transport utilization.

Mil believes that because of the production increase, use of an Mi-2 would be justified on a 12,000–20,000 hectare (30,000–50,000 acre) farm producing cereal crops and that even a farm of 5,000 hectares (12,350 acres) would produce sufficient extra yield to pay for a helicopter in one year. He also sees the helicopter playing a useful role in carrying fresh milk from pastures to markets.

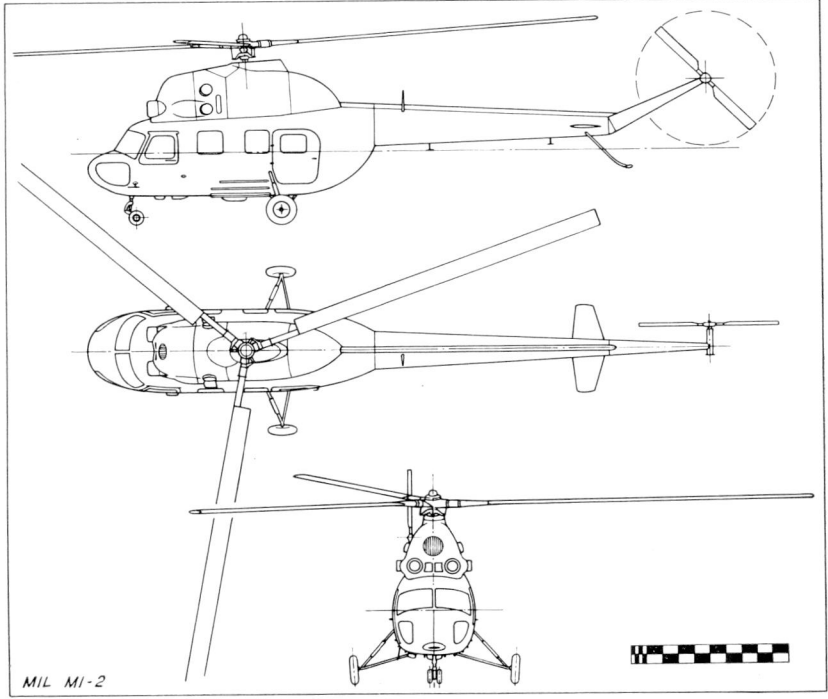

MIL MI-2

The first passenger operations by Mi-2s began in 1967 when Aeroflot's Azerbaydzhan Directorate started operating them between Baku and the Caspian offshore artificial oil island known as Neftyannye-Kamni. Introduction of these twin-turbine helicopters made it possible to carry shifts of workers in and out in much worse weather than had been possible with the single-engine Mi-4s.

The only known Soviet-registered Mi-2s are SSSR-06152, 06160, 06180 and 15834. A Polish-built Mi-2 shown at the 1967 Paris Aero Show was the dark blue, white and grey SP-PSC (c/n 530322047).

Main rotor diameter 14·5 m (47 ft 7 in); disc area 160·5 sq m (1,727·61 sq ft); length with rotors running 17·42 m (57 ft 1¾ in); fuselage length 11·94 m (39 ft 2 in); tail rotor diameter 2·7 m (8 ft 10¼ in); height to top of main rotor head 3·75 m (12 ft 3½ in); track 3·05 m (10 ft); wheelbase 2·63 m (8 ft 7½ in).

Empty weight (without operational equipment) 2,384 kg (5,255 lb) passenger and cargo versions, 2,359 kg (5,200 lb) ambulance and 2,505 kg (5,522 lb) agricultural version; maximum useful load including pilot, fuel and oil, 1,141 kg (2,515 lb); maximum chemical load 1,000 kg (2,204 lb); maximum slung load 800 kg (1,763 lb); maximum take-off weight 3,500 kg (7,716 lb).

Maximum speed at sea level 210 km/h (113·32 kt) (130·49 mph); cruising speed at sea level 205 km/h (110·61 kt) (127·37 mph); time to 1,000 m (3,280 ft) 5·5 min, to 4,000 m (13,123 ft) 26 min; service ceiling in ISA 4,000 m (13,123 ft); hover ceiling at maximum weight without ground effect 1,000 m (3,280 ft); range with 8 passengers 240 km (129 n.miles); range with supplementary tanks 590 km (318 n.miles); endurance of agricultural version at cruising speed with 5 per cent fuel reserve, 50 min with dry chemical and 40 min with liquid chemical. Spray band width 40–45 m (131–147 ft).

Mil Mi-4

Although Mikhail Mil's Mi-1 was the Soviet Union's first production helicopter, and was successful, it was small and its range of duties restricted by its limited payload. It was therefore fairly obvious that a considerably larger helicopter was required. At the end of the summer of 1951 the leading Soviet helicopter designers were called to a meeting at the Kremlin to discuss specifications for a single-engine 12-passenger helicopter and for a twin-engine type having accommodation for 24 passengers.

The outcome of the Moscow meeting was Mil's design for the Mi-4 and Yakovlev's for the Yak-24. The Yak made its first flight on 3 July, 1952, and it is believed that the Mi-4 first flew that August. Both types were powered by the 1,430/1,700 hp Shvetsov ASh-82V fourteen-cylinder two-row air-cooled radial engine, and it is thought that both types used the same type of rotor and transmission developed from those in the Mi-1, except that because of vibration troubles the Yak-24's blades were cropped. By the end of 1952 the Mi-4 had passed its State trials and been put into production.

In layout the Mi-4 is an orthodox single-rotor helicopter with tail anti-torque rotor and four-wheel non-retractable undercarriage. It has a deep front fuselage and slender tail boom. In general appearance it resembles the Sikorsky S-55 but the Mil aircraft is bigger and heavier.

Although the Mi-4 quickly passed its State trials, this does not mean that it escaped all the troubles which plagued most other early helicopters. It suffered from the usual ground resonance problems, had rotor blade flutter and short blade life. The main rotor had four tapered blades, and in the early production aircraft these had NACA-230 profile and were of mixed construction with steel-tube spar, wooden ribs and stringers and ply and fabric covering. Similar blades used on the Mi-1s were giving good service but those on the Mi-4 were 50 per cent longer and subject to much greater loadings. The life of these blades proved to be less than 100 hr, and

This Polar Mi-4, SSSR-N87, was an early aircraft and had circular windows and navigator's under-fuselage nacelle.

although much research was put into ways of improving their durability it was not until 1954 that their life was extended to 300 hr. Some six years later completely new all-metal parallel-chord blades were produced, the aft sections of honeycomb construction being bonded to the spar. This new all-metal rotor was used as standard on the 1961 model.

The Mi-4 also suffered from corrosion in magnesium fuselage components, and after 110 aircraft had been delivered magnesium parts were replaced by aluminium.

All the early production Mi-4s were military aircraft; civil interest may have been lacking because of the poor blade life, but in 1954 the 8–11 passenger Mi-4P version appeared and this began to enter Aeroflot service in considerable numbers.

Civil Mi-4s were produced for passenger and cargo transport, for ambulance work and, as the Mi-4S, for agricultural work. The transport versions were produced in very large numbers, but no evidence has been found to suggest widescale use of the type for agricultural work although soon after its appearance in 1954 an Mi-4S, SSSR-L69, was exhibited in Moscow with dry chemical distributor and liquid spray-bar. It is assumed

SSSR-31571, a standard Aeroflot Mi-4P. It was white or light grey with dark blue markings.

that agricultural work could be undertaken much more economically with An-2s and the smaller helicopters.

Although Mi-4s normally have a four-wheel land undercarriage, some with spats on all wheels, an amphibious version with four inflatable bag-type pontoons was tested in 1959 on Khimki lake to the north of Moscow.

Towards the end of 1959 an Mi-4 was delivered to Klagenfurt to the order of a syndicate of Austrian and Swiss hoteliers. This helicopter was registered SSSR-31540, described as an Mi-4S and said to be a development of the agricultural Mi-4S. It could carry up to 16 passengers and was intended to take people between Austrian and Swiss airports and clearings in the mountains and, in the off season, to be used for freight. A high-level Mil team went to Austria with the helicopter, which was the first to be sold in the West, and two Austrian pilots were selected for training, but nothing has since been heard of the venture.

Very few Mi-4s have been sold to western countries but elsewhere they have achieved satisfactory orders. By 1964 it was reported that 520 Mi-4s had been sold to some 30 countries and by late 1967 the total had reached 650. The Indian Air Force has been quoted as having as many as 80, and other military orders came from Algeria, Cuba, Czechoslovakia, Indonesia, Poland, the United Arab Republic, Yemen and Yugoslavia. ČSA's Agrolet department has at least one Mi-4, Interflug had some and so did Mongolian Airlines. Royal Nepal Airlines took delivery of two and the same number went to Yemen Airlines. One went to West Pakistan for service with the Sui Gas concern.

SSSR-31417, another standard Aeroflot Mi-4P, but without spats.

The Mi-4 was the first helicopter to enter scheduled passenger service in the USSR, which it did in November 1958 on the Simferopol–Yalta route in the Crimea. Enormous numbers of Aeroflot passengers are destined for Sochi, on the Caucasian coast of the Black Sea, and the large number of holiday resorts strung out along that coast. The nearest airport is at Adler some 30 km (18½ miles) distant, and on 1 April, 1959, regular Adler–Sochi services were begun with Mi-4s. In order to serve other resorts in the same area Mi-4 services were introduced between Adler and Gagra, Khosta, Lazarevskaya and Gelendzhik.

On 20 July, 1960, Mi-4s began operating services between the Moscow Central Air Terminal and Sheremetyevo Airport and on 1 November, 1960, began linking Bykovo and Vnukovo Airports both with Moscow and

Aeroflot's Mi-4P SSSR-66860, with spats on main wheels only, flying over the terminal building at Sheremetyevo Airport, Moscow. On the ground are the Tu-104Bs SSSR-42418 and 42434, an Li-2 and two Il-14s.

The Aeroflot Mi-4P SSSR-19125 leaving Domodedovo Airport for Vnukovo on the service which links the Moscow airports with each other and with the city.

with each other, and when the new Domodedovo Airport was opened served that as well. The only other known Mi-4 inaugural date is 11 June, 1962, when a service was begun between L'vov and Truskavets.

Mi-4s are known to have operated between Baku and Neftyannye-Kamni, the artificial island serving the oil wells in the Caspian, and there is some evidence for their operation of scheduled services between Arkhangel'sk and Severodvinsk on the White Sea coast.

Soviet sources give the impression that Mi-4s operated scheduled services in other areas, including Dagestan and the Far East, while A. M. Izakson in his *Sovetskoe Vertoletostroenie*, published in 1964, stated that whereas there were only 10 internal helicopter routes, mainly in the Crimea and Caucasian areas, being operated by Mi-4s in 1959, a year later they were operating over a hundred routes.

The Mi-4P SSSR-29079 at the Moscow city heliport. In the background is the Aeroflot terminal centre with, on the left, the Aeroflot hotel.

In addition to scheduled operations, Mi-4s were certainly used in numbers for geological and oil exploration work, for rescue and for crane duties. It is known that Mi-4s were used to erect electric transmission lines in the mountains of the Crimea and in the Kola Peninsula in the Arctic. They have been used to erect TV relay towers, carry timber from mountain forests and lay oil pipelines. They have undertaken cargo transport in many parts of the Soviet Union, being used especially in inaccessible areas in the Far East, the Northern regions, Siberia and the mountains of the Caucasus and Altai in Mongolia.

The Mi-4P SSSR-35277 was used for mail-carrying experiments in the Moscow area and made some landings on the post office roof.

Aeroflot's Mi-4P SSSR-35277 on the post office roof in Moscow. (*Aviaexport*.)

In 1967 Mi-4s are known to have been in service with Aeroflot's Azerbaydzhan, Eastern Siberia, Far East, Kazakh, Kirgiz, Northern Caucasia, Ukrainian and Yakut Directorates. They have also been operated by the Moldavian and Uzbek Directorates. Unfortunately, no statistics are available for Aeroflot's Mi-4 operations.

The cabin of the Mi-4 is 4·15 m (13 ft 7¼ in) in length, 1·78 m (5 ft 10 in) in width and 1·8 m (5 ft 11 in) in height. Volume is 16 cu m (565 cu ft). The cargo version has clamshell doors at the back of the cabin which provide an opening measuring 1·55 m (5 ft 1 in) by 1·85 m (6 ft 0¾ in), and there is a normal entrance door on the port side at the back of the cabin which measures 1·36 m (4 ft 5½ in) by 93 cm (3 ft 0½ in). The cabin is heated and soundproofed. Normal passenger accommodation is for 8–11 passengers and the cabin is provided with a wardrobe. A lavatory can be fitted but it is believed that this only applies to the eight-passenger layout.

An Indian Air Force Mi-4 flying within sight of the Himalayas.
(*Indian Air Force.*)

An economy version of the Mi-4 can have 16 seats and a de luxe variant six seats.

The cargo version can carry a 1,650 kg (3,637 lb) internal load and is capable of taking a medium-size car. Loads can be externally slung and the helicopter can have a winch capable of lifting up to 200 kg (440 lb) while hovering at up to 40 m (131 ft).

The ambulance layout can take eight stretchers, a medical attendant and oxygen equipment, and can also be arranged as an operating theatre.

The flight deck for two pilots is above the front of the cabin, and a hatch and folding steps provide communication between flight deck and cabin.

The Mi-4P has three large rectangular windows in each side of the cabin but all other versions have circular windows. Military Mi-4s have an under-fuselage navigator's compartment and this has been retained in some of the civil versions.

The ASh-82V engine has direct injection, is mounted in the extreme nose and has a two-stage supercharger. The first stage maintains nominal power up to an altitude of 1,500 m (4,921 ft) and the second stage up to 4,500 m (14,763 ft). The engine is installed at an angle of 28 deg and its transmission shaft passes through the flight deck between the crew seats. The bag-type fuel tank has 1,000 litres (219 Imp. gal) capacity and is installed above the rear of the cabin. For long-distance ferry flights a 500 litre (109 Imp. gal) tank can be installed in the cargo cabin. The main rotor blades have fluid deicing and the storage tank is immediately behind the main fuel tank. The three-blade anti-torque rotor is also protected against ice.

The main wheels are fitted with brakes and there is a tailskid just aft of the variable-incidence horizontal stabilizers.

Mi-4s are cleared for day and night all-weather operation and can be fitted with a wide range of optional radio and navigation equipment.

Latest reported overhaul lives are: rotor blades 1,500 hr, rotor head, engine and transmission 1,000 hr.

Many thousands of Mi-4s have been built and they are likely to remain in service for many years, but gradually their place will be taken in Aeroflot service by the twin-turbine powered Mi-2s and Mi-8s. Apart from Soviet production, there has been some production of Mi-4s in China.

Soviet civil Mi-4s have used a wide range of registrations but large numbers of Aeroflot Mi-4Ps have been registered in the SSSR-19000, 29000, 31000, 35000 and 66000 series. The Royal Nepal Airlines Mi-4s, which were still in service late in 1967, were 9N-HAA and 9N-HAB.

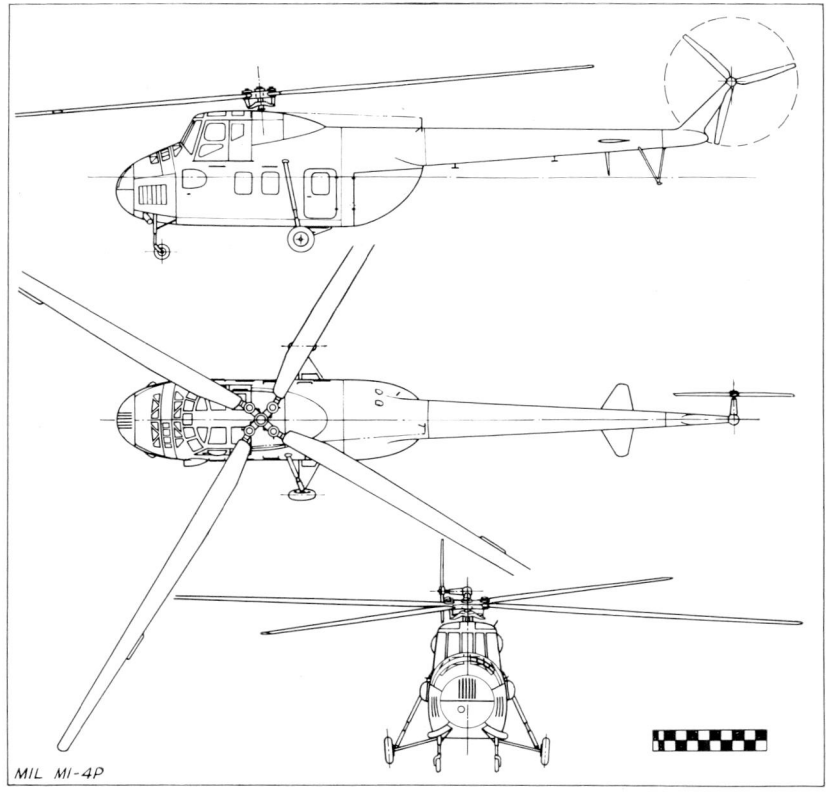

MIL MI-4P

Main rotor diameter 21 m (68 ft 11 in); disc area 346 sq m (3,724 sq ft); length with rotors running 25·02 m (82 ft 1 in); fuselage length 16·79 m (55 ft 1 in); tail rotor diameter 3·6 m (11 ft 10 in); height to centreline of tail rotor hub 4·4 m (14 ft $5\frac{1}{4}$ in); track (main wheels) 3·82 m (12 ft $6\frac{1}{2}$ in), (nosewheels) 1·53 m (5 ft); wheelbase 3·79 m (12 ft $5\frac{1}{4}$ in); ground clearance 54 cm (1 ft $9\frac{1}{4}$ in).

Empty weight 4,860 kg (10,714 lb); normal fuel 715 kg (1,576 lb); maximum ferry fuel 1,073 kg (2,365 lb); normal payload 1,485 kg (3,273 lb); maximum payload 1,835 kg (4,045 lb); normal take-off weight 7,250 kg (15,983 lb); maximum take-off weight 7,600 kg (16,755 lb).

Maximum speed at 1,500 m (4,921 ft) 210 km/h (113·32 kt) (130·49 mph); economic cruising speed 160 km/h (86·34 kt) (99·42 mph); dynamic ceiling at normal take-off weight 6,000 m (19,685 ft); dynamic ceiling at 6,000 kg (13,227 lb) take-off weight 7,200 m (23,622 ft); hover ceiling at normal take-off weight, in ground effect, 1,700 m (5,577 ft); range with normal fuel at maximum take-off weight 520 km (280 n.miles); range with supplementary fuel at maximum take-off weight 950 km (512 n.miles). Eleven passengers and 200 kg (440 lb) of baggage can be carried over a 500 km (269 n.mile) stage with 20 min fuel reserve.

The above data, supplied by Aviaexport, is for the 1961 model. The Mi-4s in service with Aeroflot appear to be mostly of earlier models with some differences in weights and performance. Weights and performance for these aircraft, where different, follow:

Empty weight 5,268 kg (11,614 lb) cargo version without nacelle, 5,356 kg (11,808 lb) cargo version with nacelle, 5,392 kg (11,887 lb) passenger version; normal fuel 750 kg (1,653 lb); maximum ferry fuel 1,140 kg (2,513 lb); payload 1,700–1,740 kg (3,747–3,836 lb); exterior slung load 1,300 kg (2,866 lb); normal take-off weight 7,350 kg (16,203 lb); maximum weight for vertical take-off 7,800 kg (17,196 lb).

Maximum flight altitude 6,000 m (19,685 ft); range with 11 passengers and 100 kg (220 lb) of baggage 250 km (135 n.miles); range with 8 passengers and 100 kg (220 lb) of baggage 400 km (215 n.miles); maximum range with 5 per cent fuel reserve 595 km (321 n.miles); maximum range without reserves 675 km (364 n.miles).

Known dates for Mi-4 introductions:

1958 November	Simferopol–Yalta (first scheduled helicopter service in USSR)
1959 April 1	Adler–Sochi
1960 July 20	Moscow City–Sheremetyevo Airport
1960 November 1	Moscow City–Vnukovo Airport–Bykovo Airport
1962 June 11	L'vov–Truskavets

Mil Mi-6

Some time during the autumn of 1957 R. I. Kaprelyan took the Mi-6 on its first flight, and on 30 October that year it carried a load of 12,004 kg (26,463 lb) to an altitude of 2,432 m (7,979 ft). When the Mi-6 first appeared it was by far the largest helicopter in the world, and 10 years later it shared this distinction with the developments of it, the Mi-10 and Mi-10K, and this trio was destined to establish numerous speed, altitude and load-carrying records.

The Mi-6 is truly enormous, having a main rotor diameter of 35 m (114 ft 10 in), a maximum weight of 42,500 kg (93,696 lb) and a maximum payload of 12,000 kg (26,455 lb).

An early Mi-6 with stub-wing and military markings. It bears the c/n 9680010V on the tail boom.

This vast Soviet helicopter was produced to meet a specification for a vehicle capable of carrying lorries, tracked vehicles, drilling rigs and other bulky loads, as well as teams of people on geological survey work in Siberia. It also had to be capable of lifting heavy loads to high altitudes in mountainous areas.

In layout the Mi-6 was fairly orthodox but because of its size new constructional techniques had to be developed, a major problem being production of rotor blades of some 17 m (55 ft $9\frac{1}{4}$ in) in length. The main rotor had five blades and was driven by two large shaft-turbines mounted on top of the fuselage. The fuselage itself was deep, with a spacious hold and a tapering tail boom which supported a four-blade anti-torque rotor. The undercarriage comprised two main units with single wheels, and twin nosewheels. There were two variable-incidence tailplanes mounted towards the rear of the tail boom and, right aft, a tailskid.

After some time the Mi-6 was fitted with a 15·3 m (50 ft $2\frac{1}{4}$ in) span wing with high angle of attack, and in cruising flight this offloaded the rotor by about 20 per cent.

It has been reported that there were five prototypes of the Mi-6 and that an initial production batch of 30 was laid down in 1960. Six were seen at the Tushino display in July 1961, and all the early examples bore military markings.

Although the early Mi-6s all had seats for up to 65 passengers, these helicopters were mainly used for the carriage of bulky loads and as cranes, but at the Soviet Exhibition in London in 1961 a model was displayed of a purely passenger version with 75 seats arranged in 15 rows of five-across. In March 1967 the Mi-6P passenger aircraft was first announced, and the prototype, SSSR-58647, was exhibited at the Paris Aero Show in May and June together with a firefighting variant.

By March 1967 more than 100 Mi-6s were in service with Aeroflot and the Soviet Air Force, and some had been delivered to the air forces of Indonesia, North Vietnam and the United Arab Republic, several of the UAR aircraft being destroyed in the Israeli–Arab war.

Mi-6s are known to have been used for carrying various kinds of vehicles including lorries and tractors, for bridge building, for carrying transmission line pylons and for a wide range of freighting and lifting. At Yaroslavl', in four days an Mi-6 lowered 16 8-ton vulcanization plants into a tyre factory through a hole in the factory roof.

In April 1966, the Mi-6 SSSR-06174 was demonstrated at Berne to Heliswiss (Schweizerische Helikopter AG) and afterwards undertook some transport work for the Swiss company. These operations took place near Herisau in Appenzell and in the mountains of the Bernese Oberland at altitudes of 500–2,800 m (1,640–9,186 ft) and in temperatures of 2–12 deg C. All loads were slung externally and weighed between 3 and $8\frac{1}{2}$ tons. At Herisau concrete masts were carried and erected for a conduit; a large wooden dome was carried from a factory at Burgdorf and placed in position on the tower of the building housing the Federal Office for Weights and Measures in Berne; components of a cable railway were flown

The Mi-6 SSSR-06174 flying at the 1965 Paris Aero Show. At that time this aircraft was operating with auxiliary fuel tanks and was painted white with grey underside, and grey and dark blue fuselage trim. It bore Aeroflot's name in white. (*Bristol Siddeley*.)

Rotor head, jet efflux pipe and wing root of the Mi-6 SSSR-06174. (*John Stroud.*)

to a mountain station in the Bernese Oberland; and a railway cabin was carried from a valley station to a mountain station at an altitude of 2,800 m (9,186 ft).

SSSR-06174 was operated without wings during its work in Switzerland but it had appeared with wings at the 1965 Paris Aero Show. This same helicopter appeared at the 1967 Paris Aero Show as a red and white painted wingless firefighting aircraft but in August 1967 crashed while fighting forest fires under difficult conditions in the south of France.

The Mi-6P passenger helicopter was still undergoing flight testing in the summer of 1967, but it is expected to go into service with Aeroflot on the services linking Moscow and the Moscow airports, as well as on other routes where its use is justified by the volume of traffic.

The first operations by Mi-6s in the Soviet Union are believed to have begun in 1961, regular cargo operations began in Turkmen on 10 August, 1963, and several years' operations are said to have proved the Mi-6 to be extremely reliable; but there is no indication of which Aeroflot Directorates, other than Turkmen, operate Mi-6s.

The Mi-6's fuselage is an all-metal semi-monocoque structure with an overall length of 33·18 m (108 ft 10¼ in). The hold is approximately square in section but is not of consistent height. The floor is 11·72 m (38 ft 5½ in) long and level for much of its length but towards the rear it curves downward. The roof slopes upward from the front, where it gives a hold height of about 2 m (6 ft 6¾ in), to a point beneath the rotor head where the height is 2·59 m (8 ft 6 in). The curve then levels off to provide a depth of 2·64 m (8 ft 8 in) above the loading sill and then descends more abruptly to finish at the rear of the large clamshell loading doors. Maximum interior width is 3·01 m (9 ft 10½ in) and floor width is 2·65 m (8 ft

8¼ in). The floor is stressed for loads of 2,000 kg/sq m (409 lb/sq ft). The clamshell doors are in two main outward-opening sections and provide absolutely unobstructed access to the hold. The lower section of each door is formed by a hydraulically-operated loading ramp attached to the rear of the hold floor. There are lashing points in the hold floor and loading is assisted by an electric winch, a cable and a hook capable of handling 800 kg (1,763 lb). A block and pulley system is supplied with each helicopter. On the side walls between each two fuselage main frames is a bench-type seat which folds flat against the wall and leaves the hold completely free of obstruction. Additional seats can be installed on the hold centreline to provide total accommodation for up to 65 passengers. With all seats installed there is still room for cargo in the aisles.

Mi-6s can be converted into ambulances capable of carrying 41 stretcher cases and two medical attendants, and have provision for carrying portable oxygen sets.

There are 1·62 m (5 ft 3¾ in) by 81 cm (2 ft 7¾ in) doors in each side at the rear of the cabin and a 1·71 m (5 ft 7¼ in) by 81 cm (2 ft 7¾ in) door forward on the port side. There are 10 circular outward-bulged windows in each side.

The Mi-6 can carry a slung load of up to 9,000 kg (19,841 lb), positioned beneath the centre of gravity, and the sling system can be installed under field conditions. This system is described in the chapter dealing with the Mi-10.

The flight deck is equipped for two pilots, radio operator and flight

View of the Mi-6 SSSR-06174 showing wing, rotor head, main undercarriage and auxiliary fuel tank. (*John Stroud.*)

The Mi-6 SSSR-06174 in Switzerland in April 1966, without wings or auxiliary fuel tanks. (*Courtesy Heliswiss.*)

engineer, with a navigator's position forward at a lower level. Full day and night all-weather instrumentation is provided, there is an autopilot, and standard equipment includes VHF and HF communications radio, radio altimeter, radio compass and directional gyros. Flight deck windows and the front windows of the navigator's cabin are electrically-heated to prevent ice formation. Storage batteries will ensure operation of vital electrical equipment for 30 min should both generators fail.

The five-blade main rotor has tapered tubular steel spars with built-up metal aerofoil sections. These aerofoil sections are each of short span and are believed to be metal-bonded to the spars. The blades have a considerable amount of droop when stationary and have fixed tabs. Controls are hydraulically actuated and all blade leading edges have electro-thermal ice protection. Main rotor speed is 120 rpm which gives a tip speed of about 792 km/h (427·36 kt) (492·12 mph).

The engines used in the Mi-6 are two Soloviev D-25V shaft-turbines which also bear the designation TV-2VM. They have a nine-stage axial-flow compressor, single-stage axial-flow turbine and an independent two-stage axial-flow free-turbine which drives the rotor through a reduction gear and shaft. The engines are mounted side-by-side and the shafts run directly back to the reduction gear but this continues to function in the case of a failure of one engine. The D-25V is 2·737 m (8 ft 11¾ in) long or with transmission shaft 5·537 m (18 ft 2 in). Dry weight, including tailpipe, is 1,200 kg (2,645 lb). Rated power is 4,700 shp and take-off power 5,500 shp. The engine will maintain take-off power up to an altitude of 3,000 m (9,842 ft) or in an ambient temperature of 40 deg C. when being run near the ground. Engine intakes and compressor guide vanes are

The Mi-6 SSSR-06174 as it appeared at the 1967 Paris Aero Show. It was painted red and white and equipped as a water-bomber firefighting aircraft. Water pipes can be seen beneath the fuselage, and the variable-incidence tailplane is in the maximum negative incidence position. In the background on the right are the Ka-25K, Mi-10K, An-24 and Il-62. On the left is Vostok. (*John Stroud.*)

provided with hot-air ice protection. Internal fuel tank capacity is 6,315 kg (13,922 lb). A cylindrical external tank of 1,745 kg (3,847 lb) capacity can be carried on each side of the fuselage just forward of the main undercarriage units, and for ferrying two more tanks of the same capacity can be installed in the cargo cabin. An automatic fuel system ensures that c.g. changes are kept to the minimum. Fuelling from ground supplies can be undertaken while hovering. The Mi-6 is equipped with a 100 hp AI-8 auxiliary power unit.

The undercarriage comprises two main single-wheel units and steerable twin nosewheels. There are low-pressure and high-pressure oleo-pneumatic chambers in the main shock struts, and the main wheels are fitted with brakes.

The Mi-6 has separate main, standby and auxiliary hydraulic systems, and in the event of a pressure drop in the main system, the standby and auxiliary systems are automatically connected. Normal use of the auxiliary system is operation of engine cowlings, loading doors and ramps, exterior sling lock and adjustment of pilots' seats.

Because Mi-6s spend much of their time operating in remote areas, careful consideration has been given to ease of maintenance and servicing. The upper parts of the engine cowlings open upward to provide access and the lower panels open downward to form working platforms. These platforms are reached through a hatch in the roof of the flight deck. Engines, main gearbox, swashplate and rotor hub can all be inspected without special ground equipment.

The Mi-6 SSSR-06174 equipped as a firefighter. (*M. J. Hooks.*)

For normal transport duties the Mi-6 operates with wings in position; when it is used as a crane the wings are generally removed, and the firefighting version, as shown in Paris, was also without wings. As a firefighter it can carry up to 14,000 kg (30,865 lb) of water, which can be either sprayed or dumped, and has a maximum take-off weight of 43,000 kg (94,799 lb), which is 500 kg (1,102 lb) greater than for other versions.

The Mi-6P resembles the standard winged version but has full airline interior for 80–100 passengers and large rectangular cabin windows.

Known civil Mi-6s are SSSR-06174 (c/n 5682010V), SSSR-11285, SSSR-11286, SSSR-11300, SSSR-11313, SSSR-11315, SSSR-11325 (c/n 4681705V), SSSR-11329, and SSSR-58647 Mi-6P.

At 1966 overhaul life of rotor blades was 700 hr, engines 500–600 hr

The Mi-6P passenger transport SSSR-58647 at the 1967 Paris Aero Show. This example, painted light grey with medium and dark green markings, is the first prototype. It was shown with 80 seats. (*John Stroud.*)

and rotor head and transmission 600–700 hr. Operating costs of the standard Mi-6 are quoted as about 60 US cents per tonne-km over stages of 150–300 km (80–161 n.miles) with 600 hr annual utilization with airframe depreciation at 6,000 hr and engine depreciation 1,800 hr.

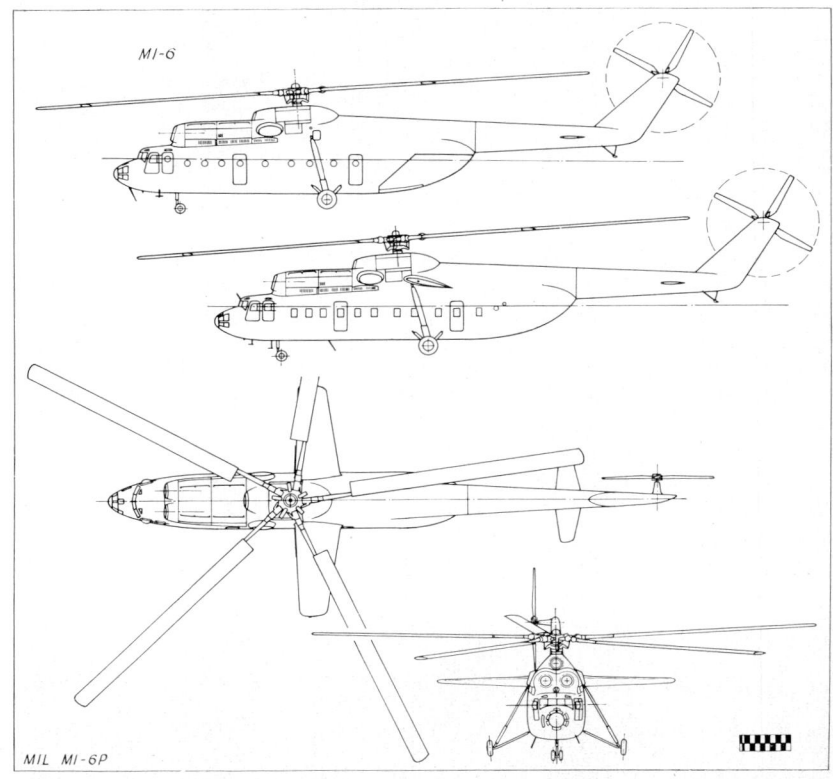

Mi-6

Main rotor diameter 35 m (114 ft 10 in); length with rotors running 41·74 m (136 ft 11½ in); fuselage length 33·18 m (108 ft 10¼ in); tail rotor diameter 6·3 m (20 ft 8 in); height to main rotor head 6·68 m (21 ft 11 in); overall height 9·86 m (32 ft 4¼ in); wing span 15·3 m (50 ft 2¼ in); span of tailplane 5·04 m (16 ft 6½ in); track 7·5 m (24 ft 7¼ in); wheelbase 9·1 m (29 ft 10¼ in); ground clearance 596 mm (1 ft 11½ in).

Empty weight 27,240 kg (60,054 lb); normal fuel 6,315 kg (13,922 lb); fuel with external tanks 9,805 kg (21,616 lb); maximum ferry fuel 13,295 kg (29,310 lb); maximum internal payload 12,000 kg (26,455 lb); maximum externally slung load 9,000 kg (19,841 lb); normal take-off weight 40,500 kg (89,287 lb); maximum take-off weight 42,500 kg (93,696 lb); maximum take-off weight with slung load, at altitudes under 1,000 m (3,280 ft), 37,500 kg (82,673 lb).

Maximum speed at normal take-off weight 300 km/h (161·88 kt) (186·41 mph) at 1,000 m (3,280 ft); maximum cruising speed at normal take-off weight 250 km/h (134·9 kt) (155·34 mph); minimum speed at 2,000 m (6,561 ft) 100 km/h (53·96 kt) (62·13 mph); dynamic ceiling at maximum weight in ISA approximately 4,400 m (14,435 ft); dynamic ceiling at maximum weight in ISA plus 30 deg C. approximately 3,300 m (10,826 ft); dynamic ceiling at 32,000 kg (70,548 lb) in ISA 6,750 m (22,145 ft); dynamic ceiling at 32,000 kg (70,548 lb) in ISA plus 30 deg C. 6,000 m (19,685 ft); static ceiling in ground effect at 40,000 kg (88,185 lb) in ISA 2,500 m (8,202 ft); static ceiling in ground effect at 40,000 kg (88,185 lb) in ISA plus 15 deg C. 2,000 m (6,561 ft); static ceiling in ground effect at 32,000 kg (70,548 lb) in ISA 4,500 m (14,763 ft); static ceiling in ground effect at 32,000 kg (70,548 lb) in ISA plus 30 deg C. 3,500 m (11,482 ft); maximum weight range with 12,000 kg (26,455 lb) payload, main tanks only and 5 per cent fuel reserve, 200 km (108 n.miles), with 9,000 kg (19,841 lb) payload 550 km (296 n.miles); maximum range at maximum weight with external tanks, 4,000 kg (8,818 lb) payload and 5 per cent fuel reserve 1,000 km (539 n.miles); normal weight range with 10,000 kg (22,046 lb) payload, main tanks only and 5 per cent fuel reserve, 225 km (121 n.miles), with 8,000 kg (17,637 lb) payload 450 km (242 n.miles); maximum range at normal weight with external tanks, 2,000 kg (4,409 lb) payload and 5 per cent fuel reserve 1,050 km (566 n.miles); maximum ferry range 1,450 km (782 n.miles). Specific fuel consumption in ISA 285 g/hp/hr (10·05 oz/hp/hr), at ISA plus 30 deg C. 295 g/hp/hr (10·4 oz/hp/hr).

Mi-6P

Dimensions as Mi-6. Empty weight 27,000 kg (59,525 lb); weight with passenger equipment, crew, oil, 20 min reserve fuel and 350 kg (771 lb) transient procedure fuel, 30,140 kg (66,447 lb); fuel and payload 12,360 kg (27,248 lb); normal take-off weight 40,500 kg (89,287 lb); maximum take-off weight 42,500 kg (93,696 lb).

Maximum speed at normal take-off weight 300 km/h (161·88 kt) (186·41 mph); cruising speed at normal take-off weight 250 km/h (134·9 kt) (155·34 mph). The Mi-6P can take-off vertically in ground effect at an elevation of 2,300 m (7,545 ft), carry 80 passengers a distance of 450 km (242 n.miles) and land vertically outside ground effect at an elevation of 2,000 m (6,561 ft). 100 passengers can be carried approximately 300 km (161 n.miles).

Mil Mi-8 (V-8)

Just as Mil's design bureau had by using turbine power transformed the Mi-1 into the far superior Mi-2, so it vastly improved the Mi-4 by designing a helicopter of very similar dimensions and powering it with turbine engines.

Work on the Mi-4 development began in 1960 and the prototype, known as the V-8, made its first flight some time in 1961. It has been reported that the V-8, later more generally known as the Mi-8, was first seen in public at the 1961 Soviet Aviation Day display, and it is known to have been demonstrated to members of the Soviet Government on 25 September, 1962.

The Mi-8 has many Mi-4 components but the use of the gas-turbine made it possible to greatly increase the cabin area. The first prototype had a four-blade main rotor driven by a single 2,700 shp Soloviev TV-2M shaft-turbine mounted above the forward end of the passenger cabin, but when the second prototype made its first flight on 17 September, 1962, it was powered by two side-by-side Izotov TV-2 shaft-turbines and, soon after, a five-blade main rotor was fitted. The prototype Mi-8s differed from subsequent models, having a door in each side of the flight deck which, like the forward passenger cabin door, opened outward.

In 1964 an Mi-8 set a speed record of 201·834 km/h (125·41 mph) over

The original Mil V-8 prototype with single Soloviev shaft-turbine and four-blade main rotor.

A prototype Mi-8 with twin-turbines and five-blade main rotor. This aircraft was later registered SSSR-06181. (*Aviaexport.*)

a 2,000 km (1,242 mile) closed circuit and also a distance record of 2,464·736 km (1,532·14 miles). An Mi-8 registered SSSR-06181 was exhibited at the 1965 Paris Aero Show, and at the 1967 Paris Aero Show the slightly modified SSSR-11052 was shown. The first single-engine prototype and the two aircraft shown in Paris all bore Aeroflot's name on their fuselages and it was stated that the Mi-8 would replace Aeroflot's Mi-4s, but it was not until late in 1967 that there came any news of the Mi-8's entry into airline service. However, towards the end of 1967 it was learned that Aeroflot's Azerbaydzhan Directorate had taken delivery of Mi-8s and put them into service between Baku, on the Caspian Sea, and the nearby oilfields where their twin-engines enabled them to provide better regularity in bad weather than had been possible with Mi-4s. It is presumed that Mi-8s will now progressively replace Mi-4s on the more important routes.

Mi-8s were seen with military markings for the first time at the big Domodedovo display in July 1967. Unlike production civil Mi-8s, these had circular windows.

The entire rear end of the Mi-8 closely resembles the Mi-4, but the placing of the engines above the cabin has given greatly increased volume for payload. The fuselage is of almost square section and provides a cabin measuring 2·34 m (7 ft 8 in) in width and 1·82 m (5 ft 11¾ in) in height. In the passenger version the floor has a length of 6·42 m (21 ft 0¾ in) and in the cargo layout is 5·34 m (17 ft 6 in) long. Rear clamshell doors provide an opening of the full width and height. Within the clamshell doors is a downward-opening passenger door of 1·7 m (5 ft 7 in) by 84 cm (2 ft 9 in). In both versions there is an aft-sliding door forward on the port side and this measures 1·405 m (4 ft 7 in) by 825 mm (1 ft 8½ in). Approximate cabin volume of the cargo Mi-8 is 23 cu m (812·24 cu ft).

SSSR-06181, a prototype twin-turbine Mi-8 at the 1965 Paris Aero Show. It was white with blue and orange-red markings. (*John Stroud*.)

A payload of 4,000 kg (8,818 lb) can be carried in the cargo hold. Ramps can be attached to the rear end of the floor to assist loading, and a 250 kg (551 lb) winch can be installed. The floor is fitted with lashing points, and to increase utilization 24 tip-up seats are provided along the cabin's side walls. Loads which are too bulky to be carried in the cabin can be carried externally provided the weight does not exceed 2,500 kg (5,511 lb). For lifting loads while hovering, a winch of 200 kg (440 lb) capacity can be fitted.

In the true passenger version there are 28 seats in seven rows, with pairs on each side of the aisle. The chairs have tip-up seats and arm rests and forward-folding backs. Aft are a wardrobe and baggage hold. The cabin is heated, or air-conditioned if required, has light luggage racks and six large rectangular windows in each side. The seats are rail-mounted and accommodation can be increased to 31–32 if required. A de luxe version has a group of six inward-facing seats along the starboard wall and, aft on the port side, a group of three seats with a table. Other amenities are a galley, lavatory and wardrobe. In the standard passenger aircraft all seats and bulkheads can be quickly removed to convert the aircraft for cargo carriage.

The passenger and cargo versions can be converted to ambulances capable of carrying 12 stretcher cases and a medical attendant.

The flight deck has normal accommodation for two pilots and there is a jump-seat for a flight engineer. The two very large bulged side windows are aft-sliding and the other nine windows provide exceptional view in most directions. The two main forward windows are equipped with windscreen wipers. The helicopter is fully instrumented for day and night all-weather operation and there is an astro-compass for flight in the Polar regions. A four-channel autopilot is fitted and incorporates altitude-locking.

The Izotov TV-2-117 shaft-turbines in the production aircraft each have a dry weight of 330 kg (727 lb), develop a nominal power of 1,200 shp and provide 1,500 shp for take-off. Full power can be maintained up to 1,600 m (5,249 ft). Specific consumption in ISA is 295 g/hp/hr (10·4 oz/hp/hr). A cooling fan, with its own air intake, is mounted above and between the engine tailpipes just ahead of the main gearbox. The engine and gearbox access panels fold down to serve as working platforms. Internal fuel capacity is 346 kg (762 lb), and there is a cylindrical external tank on each side of the fuselage to bring total fuel to 1,450 kg (3,196 lb). For ferrying, an extra fuselage tank can be fitted to increase total capacity to 2,160 kg (4,762 lb). Fuel is supplied to the engines by two booster pumps and fuel control is automatic. On production aircraft the starboard external tank

The production Mi-8 SSSR-11052 at the 1967 Paris Aero Show. Colour scheme was grey and white with dark blue trim on main fuselage and a grey band running along the tail boom. (*John Stroud.*)

fairing is extended forward to house the optional cabin air-conditioning equipment.

The main rotor shaft is inclined forward at 4 deg 30 min to the vertical, the parallel-chord rotor blades are of all-metal construction and controls are hydraulically powered. All main and tail rotor blades are protected from icing by an electro-thermal system, with ice detector and automatic or manual switching. Flight deck windows are protected by the same system and the engine intakes by hot air bled from the engine compressors.

The tail rotor is mounted on the starboard side of the tail arm and has three parallel-chord square-tipped blades. Just forward of the tail arm is a tailskid, above which are variable-incidence horizontal stabilizers.

The Mi-8's undercarriage comprises two main units with single wheels and twin steerable nosewheels.

The rotor blades are interchangeable; in case of emergency the Mi-8 can be fitted with the main rotor blades and intermediate and tail gearboxes of the Mi-4, but this would be at the expense of ice protection. Latest reported overhaul life (at 1966) for the Mi-8's engine and rotor system were: rotor blades 800–1,000 hr, engines 500–700 hr and transmission 500–700 hr.

The export price of the Mi-8 was quoted by Aviaexport in 1967 as about £250,000.

The only known Soviet-registered Mi-8s are the prototypes SSSR-06181 and SSSR-06182 and the production aircraft SSSR-11052 and SSSR-11067. It has been reported that Interflug has taken delivery of an Mi-8 registered DM-SPA. One Mi-8 has been bought by Petroleum Helicopters Inc. for use in Bolivia. This was being assembled in early 1968. Initially it will fly as N16555.

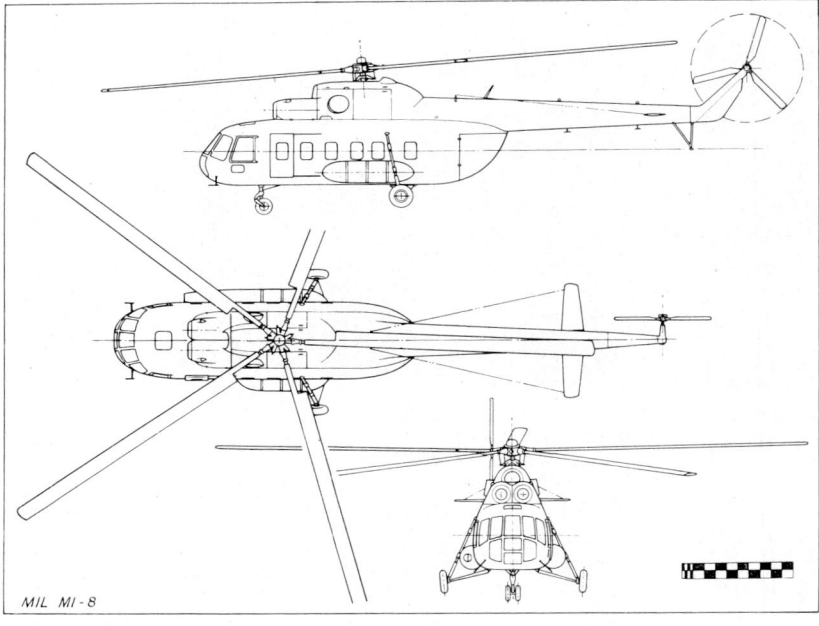

MIL MI-8

Main rotor diameter 21·29 m (69 ft 10¼ in); length with rotors running 25·28 m (82 ft 11¼ in); fuselage length 18·31 m (60 ft 1 in); tail rotor diameter 3·8 m (12 ft 5½ in); height to main rotor head 4·38 m (14 ft 4½ in); overall height 5·6 m (18 ft 4½ in); track 4·5 m (14 ft 9 in); wheelbase 4·26 m (13 ft 11¾ in); ground clearance 445 mm (1 ft 5½ in).

Empty weight, cargo version, 7,161 kg (15,787 lb), passenger version with heated cabin 7,417 kg (16,351 lb), passenger version with air-conditioning 7,509 kg (16,554 lb); normal maximum fuel 1,450 kg (3,196 lb); maximum ferry fuel 2,160 kg (4,762 lb); maximum payload 4,000 kg (8,818 lb); maximum slung load 2,500 kg (5,511 lb); normal take-off weight 11,100 kg (24,471 lb); maximum vertical take-off weight 12,000 kg (26,455 lb); maximum take-off weight with externally slung load 11,428 kg (25,194 lb).

Maximum speed at normal take-off weight, sea level to 2,000 m (6,561 ft), 230 km/h (124·11 kt) (142·91 mph); maximum speed at maximum take-off weight 220 km/h (118·71 kt) (136·7 mph); maximum speed with 2,500 kg (5,511 lb) externally slung load 180 km/h (97·13 kt) (111·85 mph); cruising speed at normal take-off weight 200 km/h (107·92 kt) (124·27 mph); ceiling at 11,200 kg (24,691 lb) take-off weight with 3,650 kg (8,047 lb) fuel and payload in ISA, 4,500 m (14,763 ft); ceiling at 10,400 kg (22,927 lb) take-off weight with 2,850 kg (6,283 lb) fuel and payload in ISA plus 25 deg C., 4,000 m (13,123 ft); range of cargo version with 2,100 kg (4,629 lb) payload and 5 per cent fuel reserve 450 km (242 n.miles); range of cargo version with 3,000 kg (6,613 lb) payload and 5 per cent fuel reserve 425 km (229 n.miles); ferry range of cargo version 650 km (350 n.miles); range of passenger version with 28 passengers, 560 kg (1,234 lb) of cargo and 30 min fuel reserve, 360 km (194 n.miles). The Mi-8 with 31 passengers can take-off vertically at sea level without ground effect, fly 300 km (161 n.miles) and land vertically at an elevation of 2,000 m (6,561 ft) without ground effect.

Mil Mi-10 (V-10) and Mi-10K

The Mil Mi-10, known also as the V-10, is a direct development of the Mi-6 and is intended mainly as a crane although capable of a wide range of duties. The Mi-10 is believed to have begun its flight trials in 1960, and it was first seen in public at the 1961 Soviet Aviation Day display at Tushino. In October 1961 an Mi-10 set an FAI-recognized record by carrying a 15,103 kg (33,296 lb) load to 2,326 m (7,631 ft), and in 1967 nine international records were still held by Mi-6 and Mi-10 series helicopters.

Like the Mi-6, the Mi-10 is powered by two 5,500 shp Soloviev D-25V shaft-turbines and it also employs the same rotor and transmission, but the fuselage and undercarriage have been completely redesigned to enable the helicopter to straddle bulky loads.

The grey and yellow Mi-10 SSSR-04102 at Le Bourget Airport, Paris, on 18 June, 1965. A 28-seat bus is attached to the cargo platform. (*John Stroud*.)

The fuselage has a straight undersurface extending from beneath the flight deck back to a point below the variable-incidence horizontal stabilizers. The engines are mounted side-by-side above the front of the main hold, and aft of the rotor the top line of the fuselage tapers quite steeply back to the fin.

Overall length of the fuselage is 32·86 m (107 ft 9½ in), and its flat underside is 3·75 m (12 ft 3½ in) above the ground. The cabin or hold area is 14·04 m (46 ft 0¾ in) long, 2·5 m (8 ft 2½ in) wide and 1·68 m (5 ft 6 in) high, and has a volume of about 60 cu m (2,118·88 cu ft). Bench-type seats along the wall provide accommodation for 28 passengers and fold against the wall when not in use, and cargo can be hoist-loaded through a door on the starboard side which measures 1·26 m (4 ft 1½ in) by 1·56 m (5 ft 1¼ in), but bulky loads are carried externally. There are 10 circular windows in the starboard side of the passenger/cargo cabin and 11 in the port. Access to the cabin and flight deck is via steps on the front port leg of the undercarriage and through an inward-opening door measuring 1·35 m (4 ft 5 in) by 78 cm (2 ft 6½ in).

The Mi-10 SSSR-04102 at Gatwick Airport on 14 March, 1967. It is carrying a larger bus than that flown at the 1965 Paris Aero Show. (*John Stroud.*)

Although the normal crew consists of two pilots and, if required, an engineer, the flight deck is spacious and equipped with five seats. Visibility is good, the forward sections of the large side windows being bulged and aft-sliding. Downward view is provided by three windows in the underside of the nose. The Mi-10 is fully equipped for day and night all-weather operation and has a gyro flux-gate compass for Polar navigation. The

Rotor head, external fuel tank and starboard loading door of the Mi-10. The loading hoist can be seen to the right of the door. (*John Stroud.*)

instrument layout is simple and well arranged, and on the centre of the panel is a television screen on which the external loads and the touchdown of the main undercarriage can be monitored by a closed-circuit system with fore and aft cameras. The flight deck windows are electrically protected against ice and there are windscreen wipers. Fans are fitted for cabin cooling.

An outstanding feature of the Mi-10 is its immense stalky quadricycle

The Mi-10 at Gatwick. This view shows the wide track of the undercarriage, generous ground clearance, droop of the rotor blades and the tailplane in the fully-up position. (*John Stroud.*)

The Mi-10 SSSR-04102 flying over Gatwick with a bus on the cargo platform. (*John Stroud.*)

undercarriage, specially designed to give maximum clearance for taxi-ing over bulky cargo loads. Each unit has twin wheels and the track of the rear wheels, between centres, is 6·92 m (22 ft 8½ in) and of the front wheels 6·01 m (19 ft 8½ in). Wheelbase is 8·29 m (27 ft 2¼ in). The main gear struts have long-stroke and oleo-pneumatic shock absorbers and the wheels are fitted with brakes. The nosewheel struts are of the levered suspension type. In spite of its great height the undercarriage allows the Mi-10 to make take-off and landing runs with forward speeds as high as 100 km/h (53·96 kt) (62·13 mph). Loading is sufficiently low to allow for safe operation from unprepared surfaces.

The Mi-10, with sling extended, moving in to pick up a truck during its Gatwick demonstration. (*John Stroud*).

The Mi-10 has three methods of carrying its main cargo loads. They can be attached by four hydraulic grips, or carried on a large platform, or they can be slung.

The cargo grips are light-weight units attached to hydraulic cylinders fixed to each undercarriage unit. By using this method of load attachment, a load measuring 20 m (65 ft 7 in) in length by 10 m (32 ft 9½ in) in width and 3 m (9 ft 10¼ in) in height can be picked up and secured in 1½–2 min. The advantage of this method over slinging is that it enables the helicopter to take-off using ground effect or with forward speed, consequently increasing the payload. The Mi-10 has been seen carrying quite large prefabricated buildings by the grip method; maximum cargo height is 3·4 m (11 ft 1¾ in).

In order to increase helicopter utilization, cargoes can be carried on pre-loaded platforms. These platforms are light metal structures measuring 8·53 m (27 ft 11¾ in) by 3·54 m (11 ft 7¼ in). They have lashing points, ramps and three wheels. Attachment points pick up the hydraulic grips.

The rear starboard load-carrying attachment of the Mi-10 with hydraulic grip holding the cargo platform to which a bus has been anchored. (*Flight International.*)

The flight deck of the Mi-10 SSSR-04102 taken while flying in the vicinity of Gatwick on 14 March, 1967. The television screen for monitoring load handling and touchdown of the main undercarriage can be seen on the centre panel. (*Flight International.*)

The maximum dimensions of loads which can be carried on the platform are 20 m (65 ft 7 in) in length, 10 m (32 ft 9½ in) in width and 3·1 m (10 ft 2 in) in height, and the maximum load is 15,000 kg (33,069 lb). An 800 kg (1,763 lb) electric winch can be installed on the platform for load handling.

For carrying slung loads there is a 1 m (3 ft 3¼ in) circular hatch in the

The hold of the Mi-10, a view taken in flight near Gatwick. The hoist and hatch can be seen in the centre. Guests occupy the tip-up seats on the starboard side and fuel tanks are on the port side. Mi-10 brochures and Mi-8 postcards occupy the table on the right. One of the external fuel tanks can be seen through the window. (*Flight International.*)

cabin floor above which can be installed a winch capable of handling an 8,000 kg (17,637 lb) load attached to a hook on the end of a sling rope. Using this method the load is attached while standing near the helicopter, the sling being hauled in to the locking point only when the helicopter is hovering. Loads of up to 500 kg (1,102 lb) can be picked up and deposited while hovering. Control of these operations is from a portable control panel attached to leads and carried in the cargo cabin.

The author has driven on to the Mi-10's platform in a 32-seat bus which was later carried by the helicopter, and the very precise control of the helicopter in hover has been observed while it handled a variety of loads.

The special short-legged Mi-10 used to establish altitude records with payload in May 1965. Although referred to as an Mi-10, in the official FAI records the designation is given as Mi-10K. Although different to the Mi-10K crane, the designation could be correct because the K stands for Korotkonogii which means short-legged.

The circular hatch in the floor provides a very fine downward view for any passengers, who are protected by a low fence. Experience of flying in the Mi-10 also showed that the helicopter is extremely stable but the proximity of the engines gives rise to considerable noise and heat. The landing of the craft, although so massive, is extremely light and delicate. When it descends towards one, the Mi-10 sounds like a giant steam locomotive.

The Mi-10 depends on hydraulic power for its control system as well as its load handling, and for safety and reliability there are main, standby and auxiliary systems. Operating pressure is 120–155 kg/sq cm (1,706·8–2,204·61 lb/sq in), and in case of a pressure drop in the main system, which supplies the boosters, the system is automatically cut out and the standby and auxiliary systems engage.

As in the case of the Mi-6, the engines maintain full power up to 3,000 m (9,842 ft) or temperatures of up to 40 deg C. There is a 100 hp AI-8 auxiliary power unit and a GSR-24 60-kW generator. Rotors, engine intakes and compressor guide vanes are all equipped with ice protection systems.

The Mi-10 brochure states that the aircraft has an internal service fuel tank of 585 kg (1,289 lb) capacity and two exterior tanks with a total capacity of 5,605 kg (12,356 lb) and that two additional cabin tanks can be installed; but other Soviet figures show normal maximum fuel capacity as 6,340 kg (13,977 lb) and maximum ferry fuel as 8,260 kg (18,210 lb).

Numbers of Mi-10s are believed to have been in service for some time with the Soviet Air Force, and pre-production civil Mi-10s have been used in eastern USSR for carrying drilling rigs, geological and engineering teams. The first Mi-10 seen outside of the Soviet Union was SSSR-04102 which was exhibited at the 1965 Paris Aero Show, and it was this Mi-10 which was demonstrated at Gatwick in March 1967. It was reported in October 1965 that series production of Mi-10s had begun, and during 1967 the civil aircraft SSSR-04103 and SSSR-04105 were seen in the Soviet Union.

In 1967 Petroleum Helicopters Inc. of Lafayette, Louisiana, assembled an Mi-10 from imported components. This helicopter was registered N16556 and was acquired for use in Bolivia.

The price of the Mi-10 was quoted by Aviaexport in 1967 as £500,000–600,000. Operating costs have been published as about 70 US cents per tonne-km over stages of 100–200 km (54–108 n.miles), with aircraft-km costs of $9 over a 100 km (54 n.mile) stage and $8 over stages of about 200–300 km (108–161 n.miles). These figures assume 600 hr annual

Close-up of the record-breaking short-legged Mi-10 which carried a 25,000 kg (55,116 lb) load to 2,840 m (9,317 ft). The pilots were G. Alferov and G. Karapetyan and engineers N. Genov and V. Shein.

The prototype Mi-10K with short undercarriage legs and under-nose cabin from which the helicopter can be controlled while serving as a crane.

utilization, with 6,000 hr airframe depreciation and 1,800 hr engine depreciation. Engines, rotor blades, rotor head and transmission are believed to have the same overhaul lives as those in the Mi-6.

In 1965 a specially modified Mi-10 with short-legged light-weight tricycle undercarriage, having spats on the main wheels, established two records by lifting a 5,175 kg (11, 408 lb) load to 7,134 m (23,405 ft), and a 25,105 kg (55,347 lb) load to 2,800 m (9,186 ft).

A new version of the Mi-10 was produced in 1965 and exhibited in Moscow on 26 March, 1966. This is the Mi-10K (K for Korotkonogii—short-legged), which rather naturally is distinguished by its short undercarriage legs. The fuselage, engines and rotor system are identical to those of the Mi-10 but its stalky quadricycle undercarriage has been replaced by shorter main legs and simple nose units with small twin wheels. The

The Mi-10K SSSR-29115 flying at the Paris Aero Show in 1967. (*Jean Alexander.*)

The forward fuselage and engines of the Mi-10K SSSR-29115. The under-nose control cabin can be clearly seen. This example was painted white with light and dark grey trim. (*John Stroud.*)

Mi-10K also has a small cabin for a pilot beneath the nose. This has a backward-facing seat and controls for handling the aircraft in hover and for operating the hoist. This version of the Mi-10 is intended mainly for construction and engineering work requiring a crane rather than for more normal cargo carriage. The maximum permissible slung load is 11,000 kg (24,251 lb) and this is expected to be increased to 14,000 kg (30,865 lb)

Under-nose control cabin of the Mi-10K, showing backward-facing seat, control column, instrument panel and cooling fan. (*John Stroud.*)

Main undercarriage, external fuel tank and rotor head of the Mi-10K. (*John Stroud.*)

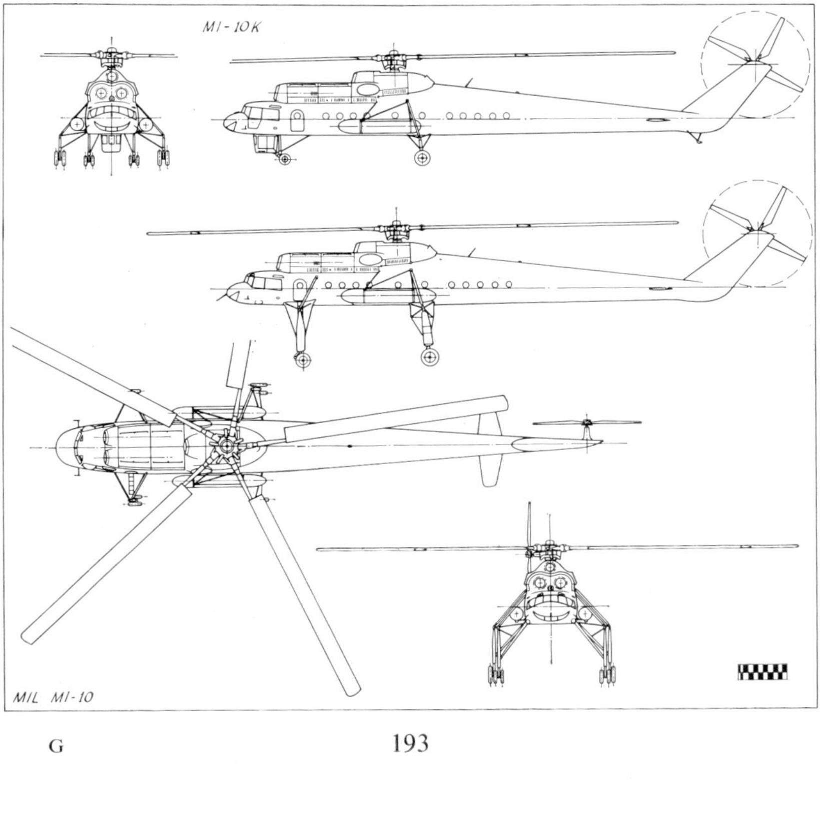

when the Soloviev D-25V engines are uprated to 6,500 shp. Fuel capacity in internal and external tanks is 9,000 litres (1,979 Imp. gal).

It was stated in mid-1967 that three Mi-10Ks were engaged in flight testing, but the only known aircraft is SSSR-29115 (c/n 5680156) which appeared at the 1967 Paris Aero Show, bearing Aeroflot's name on its external fuel tanks.

Records held by the Mi-10K in 1967 appear in Appendix V.

Mi-10

Main rotor diameter 35 m (114 ft 10 in); length with rotors running 41·89 m (137 ft 5¼ in); fuselage length 32·86 m (107 ft 9½ in); tail rotor diameter 6·3 m (20 ft 8 in); height to main rotor head 8·45 m (27 ft 8½ in); overall height 9·9 m (32 ft 5¾ in); track (main undercarriage outer wheels) 7·55 m (24 ft 9¼ in), (outer nosewheels) 6·4 m (21 ft); wheelbase 8·29 m (27 ft 2¼ in); fuselage ground clearance 3·75 m (12 ft 3½ in).

Empty weight 27,000 kg (59,525 lb); normal maximum fuel 6,340 kg (13,977 lb); maximum ferry fuel 8,260 kg (18,210 lb); maximum payload when carried by hydraulic grips or on cargo platform (including platform weight) 15,000 kg (33,069 lb); maximum payload carried on external sling 8,000 kg (17,637 lb); normal vertical take-off weight 43,450 kg (95,791 lb); maximum vertical take-off weight with slung load 38,000 kg (83,776 lb).

Maximum speed at normal take-off weight at sea level to 1,000 m (3,280 ft) 200 km/h (107·92 kt) (124·27 mph); maximum speed for ferry flight at 38,000 kg (83,776 lb) take-off weight at sea level to 1,000 m (3,280 ft), 235 km/h (126·8 kt) (146 mph); cruising speed at normal take-off weight 180 km/h (97·13 kt) (111·85 mph); service ceiling at 42,000 kg (92,594 lb) in ISA, 4,000 m (13,123 ft) with 14,500 kg (31,967 lb) fuel and payload; service ceiling at 42,000 kg (92,594 lb) in ISA plus 30 deg C., 3,000 m (9,842 ft) with 14,500 kg (31,967 lb) fuel and payload; service ceiling at 36,000 kg (79,366 lb) in ISA, 5,500 m (18,044 ft) with 8,500 kg (18,739 lb) fuel and payload; service ceiling at 36,000 kg (79,366 lb) in ISA plus 30 deg C., 4,700 m (15,419 ft) with 8,500 kg (18,739 lb) fuel and payload; range at normal take-off weight with maximum payload about 30 km (16 n.miles); range at normal take-off weight with 12,000 kg (26,455 lb) payload 250 km (135 n.miles); range at normal take-off weight with 10,000 kg (22,046 lb) payload 400 km (215 n.miles); maximum ferry range 630 km (340 n.miles). Specific fuel consumption in ISA 290 g/hp/hr (10·22 oz/hp/hr).

Maximum flight altitude for service is limited to 3,000 m (9,842 ft). Range figures include 5 per cent fuel reserve.

Mi-10K

Dimensions, except heights, as Mi-10.

Empty weight 24,680 kg (54,410 lb); maximum payload 11,000 kg (24,251 lb); maximum useful load (fuel and payload) 12,170 kg (26,829 lb); maximum slung load 11,000 kg (24,251 lb); take-off weight 38,000 kg (83,776 lb).

Cruising speed without payload 250 km/h (134·9 kt) (155·34 mph); cruising speed with externally slung load 200 km/h (107·92 kt) (124·27 mph); absolute ceiling with 12,170 kg (26,829 lb) useful load in ISA 5,000 m (16,404 ft); hover ceiling in ground effect ISA, 4,000 m (13,123 ft) with 9,000 kg (19,841 lb) useful load; a payload of 10,000 kg (22,046 lb) can be carried 70 km (37 n.miles) at 2,000 m (6,561 ft) cruise level and landed at an elevation of 2,500 m (8,202 ft); range at maximum take-off weight with 8,000 kg (17,637 lb) useful load and 5 per cent fuel reserve 300 km (161 n.miles).

Tupolev Tu-70

In 1944 four USAAF Boeing B-29s, engaged on bombing raids on Japan from bases in China, made forced landings on Soviet territory and were impounded, the Soviet Union not then being at war with Japan. At that time the B-29 was a much more advanced design than any Soviet bomber, and after a thorough study of the American aircraft the USSR began production of a copy of it. This was the Tupolev Tu-4 which has been reported as being in series production by March 1945. Dimensions of the Tu-4 were almost certainly the same as those of the B-29; in place of the B-29's 2,200 hp Wright R-3350 engines the Tu-4 is believed to have had four Shvetsov ASh-90 engines each developing 2,350 hp for take-off.

The Tu-70 transport prototype developed from the Tu-4 bomber, itself a copy of the Boeing B-29.

From the Tu-4 the Soviet Union developed a commercial transport with the designation Tu-70. The Tu-70 was first seen at Tushino near Moscow on 3 August, 1947, and was found to be essentially a Tu-4 with a new and longer fuselage providing accommodation for 72 passengers and six crew. This prototype had a red star on its fin and rudder, was without a registration or any form of identification, and the type was not put into production, Aeroflot having no interest in it.

In layout the Tu-70 was a low-wing cantilever monoplane with single fin and rudder, four radial air-cooled engines and retractable nosewheel undercarriage.

The wing is thought to have been identical to that of the B-29, which was a two-spar metal structure with fabric-covered ailerons. Aspect ratio was 11·58 and there was taper in thickness and plan although the trailing edge was almost untapered. Fowler-type flaps added 20 per cent to the wing area when extended for landing. Five groups of bladder-type fuel cells provided a total capacity of 29,450 litres (6,478 Imp. gal). A thermal deicing system protected the leading edge. The tail unit was of similar construction to the wing, and had fabric-covered control surfaces which were aerodynamically and mass balanced. Thermal leading edge ice protection was provided.

Close-up of the Tu-70 showing its close resemblance to the Boeing B-29.

The Tu-70's fuselage was of circular section and may have provided the Soviet Union's first exercise in design and production of a large pressurized fuselage. Complete details of the interior layout are not known, but the train-type luggage racks, bulkhead windows and use of lace decoration was to be repeated in the early Tu-104s nearly 10 years later. Seating in the Tu-70 was in reclinable pairs on each side of the central aisle, and the lace antimacassars bore five-pointed stars and hammers and sickles. The extreme nose of the aircraft housed the navigator and incorporated the transparent bomber nose of the B-29 design.

Judging from the window arrangement there would appear to have been a main cabin aft, with seven circular windows each side, forward of this a small cabin with two quite large rectangular windows each side, then came a buffet section with two circular windows in the ceiling on each side some way from the centreline. Forward of the buffet there appear to have been two more cabins each with two rectangular windows in each side; and between these forward cabins circular windows high up suggest there were lavatories. Sleeping berths could be provided. The entrance door was aft on the port side.

The undercarriage comprised the main gear with twin wheels which retracted forward into the inboard engine nacelles, and twin nosewheels which were backward retracting. There was a retractable tail bumper.

The ASh-90 engines were copies of the Wright R-3350 eighteen-cylinder two-row air-cooled radials used in B-29s. The deep cowlings with separate lower position air intakes suggest that the oil coolers and supercharger inter-cooler occupied the same positions as on the R-3350. The four-blade airscrews are believed to have been based on Hamilton Standard design, with constant-speed, and fully-feathering and braking ability.

The Tu-70 taking-off with flaps extended and retractable tail bumper lowered.

In an air display at Tushino in 1951 the Tu-75 military version of the Tu-70 was seen. This had a further lengthened fuselage, large doors fore and aft and accommodation for 100 armed troops or a smaller number of paratroops. It is not known whether the military aircraft was a modification of the civil prototype or a new aircraft.

Span 43·08 m (141 ft 4 in); length 36·29 m (119 ft 0¾ in); height 9·15 m (30 ft 0¼ in); wing area 161·56 sq m* (1,739 sq ft). Weights are unknown but maximum take-off weight is likely to have been about 65,000–70,000 kg (143,300–154,324 lb). Maximum speed at 9,000 m (29,527 ft) 515 km/h (277·89 kt) (320 mph); cruising speed 370 km/h (199·65 kt) (229·91 mph); maximum range 4,800 km (2,590 n.miles).

All figures should be regarded as approximate.

* Wing area quoted is that of the Boeing B-29 which was almost certainly the same.

Part of the main cabin of the Tu-70. View looking forward.

The prototype Tu-104, SSSR-L5400, at London Airport (Heathrow) in March 1956. The fuselage paint lines were blue with white outline and a thin red line beneath the lower stripe. The Tupolev monogram was red on a white ground with blue outline.
de Havilland Aircraft Co.)

Tupolev Tu-104

In 1952 Aeroflot is believed to have carried about 2 mn passengers and 200,000 tons of mail and freight, and three years later these figures rose to 2,500,000 passengers and 258,700 tons of mail and freight. During those three years this volume of traffic was handled by a fleet which had as its largest and fastest units the Li-2 and Il-12, although it was at the end of November 1954 that the Il-14 was introduced into regular passenger service.

The rapidly growing volume of passenger and cargo traffic combined with the considerable distances to be covered within the Soviet Union made it vital for Aeroflot to be re-equipped with much larger aircraft of greatly improved performance.

In 1953, as part of a major programme to modernize Aeroflot, design began of the Soviet Union's first turbojet transport. This was the Tu-104 designed by a team led by A. N. Tupolev and construction began in 1954. The Tu-104 was an enormous advance on all previous Soviet transport aircraft, and to achieve production in the shortest time it was designed to incorporate the wings, tail unit, undercarriage, engine installation and fuselage nose of the Tu-16 twin-turbojet bomber, several of which were seen in public for the first time in 1954.

By using standard components on a large scale the Tu-104's production time was cut to the minimum, and on 17 June, 1955, the prototype, SSSR-L5400, flown by test pilot Grozdov, made its first flight at Vnukovo Airport, Moscow.

The development aircraft SSSR-L5402 is said to have appeared at the Aviation Day display at Moscow's Tushino Airport on 3 July, 1955, but outside the USSR nothing was known of the appearance of the Tu-104 until on 22 March, 1956, SSSR-L5400 arrived at London Airport (Heathrow) on a special flight. As far as is known the first descriptions of the aircraft did not appear in the Soviet press until 25 March, 1956, that is, three days after its arrival in London. A little later in the year three Tu-104s arrived at Heathrow, they were SSSR-L5400, L5412 and L5413 and bore on their noses the unexplained numbers 25, 29 and 27.

Crew training on Tu-104s began in May 1956, with the Tu-104G which was for some reason known to the training unit as *Krasnaya Schapochka* (*Little Red Riding Hood*).

The early batch of Tu-104s made numerous proving flights both in and outside the USSR, and on 15 September, 1956, the type went into regular service over the Moscow–Omsk–Irkutsk route. At the time of its entry into service the Tu-104 was the only turbojet-powered transport in airline service, the de Havilland Comet 1 and 1A types having been withdrawn from service in 1954. It was not until the autumn of 1958 that BOAC introduced Comet 4s and Pan American World Airways the Boeing 707-120s.

The Tu-104 was a low-wing monoplane with sweptback wing and tail surfaces and had a most attractive appearance. Even now, after more than a decade of service, the Tu-104 looks modern and businesslike.

The wing, which had anhedral, was a two-spar structure with 12 per cent thickness/chord ratio, 35 deg sweepback at 25 per cent chord and 40 deg 30 min leading edge sweep inboard and 37 deg 30 min outboard. Fowler-type slotted flaps extended from the undercarriage fairings to the ailerons,

The prototype Tu-104 at London Airport in March 1956. This view shows the anhedral wing, close-in engine layout and main undercarriage housings. Immediately forward of the nosewheel doors is the chin radar. (*de Havilland Aircraft Co.*)

The prototype Tu-104, SSSR-L5400, and production aircraft SSSR-L5412 (right) and SSSR-L5413 at London Airport (Heathrow) in the spring of 1956. (*Flight International.*)

and short-span flaps occupied the trailing edge between the undercarriage fairings and the engine tailpipes. There were two boundary layer fences on the upper surface of each wing. Hot air from the engine compressors provided ice protection for the leading edge and this was vented from two vertical openings in each wing tip. The ailerons had inset trim tabs, and a static-discharge rod was attached to the trailing edge of each aileron at its inboard end with another near the undercarriage fairing. Fuel was in bag tanks in the outer wings and in the fuselage, with capacity believed to have been 27,215 kg (59,999 lb), and provision was made for fuel jettisoning.

The fuselage was a circular-section semi-monocoque structure pressurized to a differential of 0·5 kg/sq cm (7·1 lb/sq in) and having a pressure bulkhead between the forward passenger cabin and the crew compartment.

Fin and rudder and tailplane and elevators were all-metal structures with 45 deg sweepback. Controls were manually operated but the rudder had hydraulic boost. Tabs were fitted to the rudder and both elevators. Electro-thermal leading edge deicing was provided.

The main undercarriage units comprised four-wheel bogies which were hydraulically-retracted backward into streamlined fairings which protruded aft of the wing trailing edge. Anti-skid brakes were fitted. The steerable twin nosewheels were backward retracting and enclosed in flight by two doors. Under the rear fuselage was a retractable bumper

SSSR-L5423, one of the early Tu-104As, taking-off from Le Bourget Airport, Paris. (*Aeroplane*)

and aft of this a compartment housing two braking parachutes. Mainwheel tyre pressure was 9 kg/sq cm (128 lb/sq in) and nosewheel tyre pressure 7·5 kg/sq cm (106·67 lb/sq in).

The two 6,750 kg (14,881 lb) thrust Mikulin RD-3 or AM-3 eight-stage axial-flow turbojets were carried by the fuselage structure and buried in the wing roots. Neither noise suppressors nor thrust reversers were fitted, but engine intakes were equipped with deicing.

The Tu-104 had accommodation for 50 passengers. Aft of the forward

The Tu-104A SSSR-42382 at London Airport (Heathrow) on 7 May, 1959, after a proving flight from Moscow. Boundary layer fences, undercarriage fairings and galley roof windows are all visible. (*Flight International.*)

entrance were two small cabins, the forward one with pairs of facing seats separated by a table on the starboard side and two seats on the port side. These single chairs were positioned at an angle to give almost sideways seating. The aft of the two forward cabins had facing pairs of seats on each side and these were separated by tables. At a higher level over the spars was the buffet section and this incorporated some glass-fronted cabinets containing porcelain figurines. Immediately aft of the buffet was an eight-seat cabin with facing pairs of seats on each side and then came the main cabin with 28 seats arranged in forward-facing pairs on each side of the centre aisle. Right aft was a large area providing a rear entrance vestibule, two coat cupboards, a baggage compartment and two large lavatories and wash rooms. A third lavatory was situated on the starboard side opposite the forward entrance door.

Passenger cabin dimensions were: 16·11 m (52 ft $10\frac{1}{4}$ in) long, 3·2 m (10 ft 6 in) wide and 1·95 m (6 ft $4\frac{3}{4}$ in) high. Volume excluding the flight deck was 142·3 cu m (5,025·28 cu ft).

The prototype Tu-104 did not appear to have underfloor baggage holds; but the production Tu-104s had an underfloor hold forward of the spars, four underfloor holds aft of the spars and a cabin baggage area opposite the rear entrance. These holds were numbered 1 to 6 from front to rear. The underfloor holds were served by three inward-opening hatches. The largest hold was No. 1 which was 2·5 m (8 ft 2½ in) long, 1·6 m (5 ft 3 in) wide and 80 cm (2 ft 7½ in) high. Its volume was 3·2 cu m (113 cu ft). Volume of the other holds was 2·6 cu m (72·74 cu ft) No. 2, 0·8 cu m (28·25 cu ft) No. 3, 1·4 cu m (49·44 cu ft) No. 4, 2 cu m (70·63 cu ft) No. 5 and 2·9 cu m (102·41 cu ft) No. 6. Maximum capacity of the holds was 960 kg (2,116 lb) No. 1, 780 kg (1,719 lb) No. 2, 210 kg (462 lb) No. 3, 450 kg (992 lb) No. 4 and 600 kg (1,322 lb) No. 5. No. 6 hold was not to be used for cargo but could take 350 kg (771 lb) of baggage or 360 kg (793 lb) of mail. Maximum permissible floor loading was 400 kg (881 lb) per square metre in the underfloor holds and 200 kg (440 lb) in No. 6 hold. The underfloor hatches were all 78 cm (2 ft 6·7 in) wide and 1·12 to 1·3 m (3 ft 8 in to 4 ft 3 in) long. Passenger entrance doors, which opened inward, were 1·3 m (4 ft 3 in) high and 70 cm (2 ft 3½ in) wide.

The Tu-104's flight deck was somewhat unorthodox by European and American standards. The two pilots were separated by a passageway to the navigator's position at a lower level in the extreme nose. Radio operator's and flight engineer's seats were at the rear of the control cabin. The nose itself was identical to that of the Tu-16 bomber, incorporated an optically-flat panel in the underside and also had chin radar. In the flight deck and navigator's cabin were no less than 22 separate windows. The pilots' main windows were provided with deicing and de-misting, and cabin cooling was provided by small fans—a feature of most Soviet flight decks.

A line-up of Aeroflot Tu-104s at Vnukovo Airport, Moscow, on 2 April, 1960. The nearest aircraft, SSSR-42345, is a Tu-104A. (*British European Airways.*)

The flight deck of the Tu-104A SSSR-42382. The navigator's compartment is in the nose at a lower level. (*Flight International.*)

The passenger cabin windows were circular, three of them in the port side and four in the starboard were in emergency exits. There were seven windows in each side of the main cabin and two in each side of the three small cabins. In the buffet there were two windows in the starboard side, these were slightly higher than the other cabin windows, and there were three windows in the roof to port of the centreline. On the starboard side there were four windows aft of the passenger cabins, one in the forward lavatory and one beside the radio operator. On the port side there were two lavatory windows right aft, and one window in the forward entrance door.

It is thought that only a small batch of Tu-104s was built, possibly 20 registered SSSR-L5401–SSSR-L5420, and although the type made a number of long-distance flights in 1956, including one that October to Tashkent, Delhi and Rangoon, the only routes they were known to have served that year were the Moscow–Omsk–Irkutsk and, from 12 October, the Moscow–Prague route.

Although the Tu-104 had an impressive performance, cruising at about

806 km/h (435 kt) (500 mph), it was not economic to use 13,500 kg (29,762 lb) of thrust to carry only 50 passengers, and the decision was taken very early to produce a 70-passenger version. This second version of the Tu-104 was the Tu-104A which was first shown in Moscow in July 1957.

It is thought that the first Tu-104A was SSSR-L5421 and it was this aircraft which on 6, 11 and 24 September, 1957, set a number of official speed and load-carrying records for turbojet aircraft. On 6 September

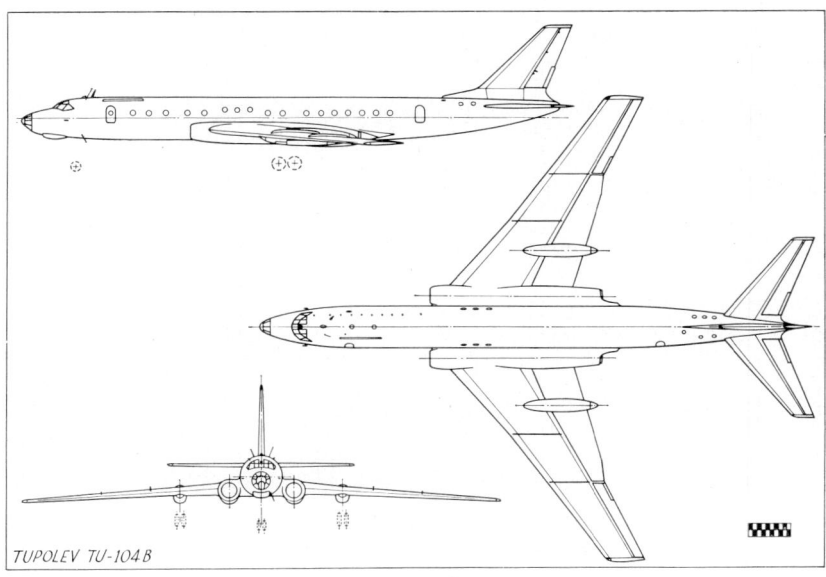

TUPOLEV TU-104B

this Tu-104A captured the record for the greatest load carried to 2,000 m (6,561 ft) when it flew from Vnukovo with 20,053 kg (44,208 lb). On 11 September SSSR-L5421 set a speed record of 897·498 km/h (557·67 mph) over a closed circuit of 2,000 km (1,242 miles). The load carried over the circuit brought with it the speed record over 2,000 km (1,242 miles) with 1,000 kg and 2,000 kg (2,204 lb and 4,409 lb) loads. On 24 September the Tu-104A set a speed record of 970·821 km/h (603·29 mph) over a 1,000 km (621 miles) circuit and at the same time secured the 1,000 km (621 miles) record for loads of 1,000, 2,000, 5,000 and 10,000 kg (2,204, 4,409, 11,023 and 22,046 lb). On 6 September the aircraft had set an altitude record of 11,221 m (36,814 ft).

The Tu-104A differed from the Tu-104 in having two Mikulin AM-3M turbojets, each developing 8,700 kg (19,180 lb) take-off thrust, and in its interior layout. Passenger accommodation was divided into two cabins, a forward cabin with 16 rail-mounted first class seats arranged in forward-facing pairs on each side, and a main cabin aft of the spars with 54 seats

Main cabin of the Tu-104A SSSR-42382, view looking aft. (*Aeroplane*.)

arranged in 10 rows of triple units on the starboard side and with double units to port. At the front of the cabin were two aft-facing seats on each side. First class seat pitch was 1,110 mm (43·7 in) and tourist class 930–960 mm (36·6–37·8 in). The buffet area was reduced in size and the general furnishing had a less Victorian appearance than the Tu-104. As in the earlier aircraft there was an emergency oxygen supply to each seat. The underfloor holds were slightly modified but of approximately the same

Facing seats at the front of the Tu-104A's main cabin. Passenger oxygen supply equipment can be seen beneath the left-hand window. (*Aeroplane*.)

capacity, but a hold of 3 cu m (105·94 cu ft) on the starboard side between the flight deck and the front cabin replaced the No. 6 hold of the Tu-104.

Total fuel capacity of the Tu-104A is believed to have been the same as that of the Tu-104 and was carried in a group of flexible tanks in the fuselage and three groups of tanks in each outer wing. Total usable fuel was 33,150 litres (7,292 Imp. gal).

The Tu-104A entered service with Aeroflot during 1957 and from that time it has not always been possible to distinguish between operations by Tu-104s and Tu-104As.

On 12 June, 1957, the Tu-104 type began operating over the Moscow–Sverdlovsk–Novosibirsk route but it is not known which model was used;

Tu-104A SSSR-L5429. Of particular interest is the Tu-104 on the left; this bears a red star on the fin and is the only military Tu-104 known to have appeared in a photograph.

however, it is known that Tu-104As began operating Moscow–Tashkent–Delhi services on 18 August, 1958, and Moscow–Tirana–Cairo services on 5 December, 1958. By the winter 1958–59 Tu-104As were also working on the following routes: Moscow–Omsk–Irkutsk–Peking, Moscow–Paris, Moscow–Prague, Moscow–Brussels, Moscow–Amsterdam, Moscow–Copenhagen and Moscow–Budapest.

In 1959 the Tu-104A was introduced on a number of routes including Moscow–Irkutsk–Chita–Pyongyang (23 March), Moscow–Copenhagen–London (16 May) and Leningrad–Sverdlovsk–Novosibirsk–Irkutsk–Vladivostok (21 May). Tu-104s are reported as going into service on the Moscow–Vladivostok route on 8 January, the Moscow–Petropavlovsk-Kamchatskiy and Moscow–Kiev routes on 15 May and Moscow–Khabarovsk route on 1 August. It is not known whether these last-mentioned operations were by Tu-104s or Tu-104As.

The winter of 1960–61 saw Tu-104As working Moscow–Kiev–Vienna services, and in 1962 they began operating a Moscow–Djakarta service (31 January) and the Moscow–Berlin services (4 April).

The Soviet Union made efforts to export the Tu-104 series but only six were sold. These were Tu-104As bought by ČSA—Československé Aerolinie. The first ČSA Tu-104A, OK-LDA *Praha* (*Prague*), made its first service flight, from Prague to Moscow, on 15 November, 1957. OK-LDB *Bratislava* and OK-LDC *Brno* were in service by the end of

Tu-104B

Span 34·54 m (113 ft 4 in); length 40·06 m (131 ft 5 in); height to top of tail 11·9 m (39 ft 0½ in); wing area 183·5 sq m (1,975·17 sq ft); aspect ratio 6·5; sweepback at 25 per cent chord 35 deg; track 11·83 m (38 ft 10 in); wheelbase 15·32 m (50 ft 3 in).

Empty weight 42,500 kg (93,696 lb); maximum fuel 26,500 kg (58,422 lb); maximum payload 12,000 kg (26,455 lb); normal take-off weight 76,000 kg (167,551 lb); normal landing weight 61,000 kg (134,482 lb); maximum landing weight 64,000 kg (141,096 lb).

Maximum speed at 10,000 m (32,808 ft) 950 km/h (512·62 kt) (590·3 mph); cruising speed 750–800 km/h (404·7–431·68 kt) (466–497·1 mph).

Take-off run at 76,000 kg (167,551 lb) 2,200 m (7,217 ft); landing run at 61,000 kg (134,482 lb) 1,850 m (6,069 ft).

Service ceiling 11,500 m (37,729 ft).

Maximum payload range at 10,000 m (32,808 ft) against 50 km/h (26·9 kt) headwind with 1 hr fuel reserve, 2,100 km (1,133 n.miles); range with 6,000 kg (13,227 lb) payload and same allowances, 3,100 km (1,672 n.miles).

Fuel consumption at 780 km/h (420·89 kt) (484·67 mph) at 10,000 m (32,808 ft), 2,400 kg (5,291 lb) per engine per hour.

The original Tu-104 had the same dimensions as the Tu-104A. Maximum payload was 5,200 kg (11,464 lb), maximum take-off weight 71,000 kg (156,528 lb) and maximum payload range with 1 hr fuel reserve 2,650 km (1,430 n.miles).

The Tu-104A SSSR-42382 using braking parachutes when landing on a snow-covered runway at Schiphol Airport, Amsterdam, in January 1959.

Known dates for Tu-104 introductions and technical proving flights:

Aeroflot

1956 March 22	Moscow–London (special flight by SSSR-L5400)
1956 September 15	Moscow–Omsk–Irkutsk (first Tu-104 service)
1956 October 12	Moscow–Prague (Tu-104)
1956 October 18	Moscow–Tashkent–Delhi–Rangoon (Tu-104—technical flight)
1957 June 12	Moscow–Sverdlovsk–Novosibirsk
1957 October 24	Moscow–Petropavlovsk-Kamchatskiy (technical flight)
1958 August 18	Moscow–Tashkent–Delhi (Tu-104A)
1958 December 5	Moscow–Tirana–Cairo (Tu-104A)
1959 January 8	Moscow–Vladivostok
1959 March 23	Moscow–Irkutsk–Chita–Pyongyang (Tu-104A)
1959 April 15	Moscow–Leningrad (first Tu-104B service)

1959 May 15	Moscow–Petropavlovsk-Kamchatskiy and Moscow–Kiev
1959 May 16	Moscow–Copenhagen–London (Tu-104A)
1959 May 21	Leningrad–Sverdlovsk–Novosibirsk–Irkutsk–Vladivostok (Tu-104A)
1959 June 15	Irkutsk–Yakutsk
1959 July 7	Moscow–Kiev (Tu-104B)
1959 August 1	Moscow–Sverdlovsk–Novosibirsk–Khabarovsk
1960 February 10	Moscow–Simferopol (Tu-104B)
1960–61 winter	Moscow–Kiev–Vienna (Tu-104A)
1961 May 15	Moscow–Kuybyshev
1961 May 15	Moscow–Mineral'nyye Vody (Tu-104B)
1962 January 31	Moscow–Tashkent–Delhi–Rangoon–Djakarta (Tu-104A)
1962 April 4	Moscow–Berlin (Tu-104A)
1963 summer	Moscow–Stockholm (Tu-104A)
1963 December 1	Moscow–Chelyabinsk
1964 May 15	Moscow–Kutaisi
1964 June 15	Moscow–Petropavlovsk-Kamchatskiy–Vladivostok
1964 November 15	Moscow–Sukhumi and Moscow–Ul'yanovsk (Tu-104B)
1965 April 15	Moscow–Odessa (first Tu-104D service)
1965 summer	Odessa–Leningrad (Tu-104D)
1965 summer	Kiev–Prague (Tu-104B)
1965 October 12	Chita–Moscow
1967 July 31	Moscow–Zürich (Tu-104A)
1968 April 3	Leningrad–Stockholm–Amsterdam (Tu-104A)
1968 April 5	Leningrad–Copenhagen–London (Tu-104B)

ČSA

1957 November 15	Prague–Moscow (first ČSA Tu-104A service, by OK-LDA)
1960 April 1	Prague–London (first service by OK-NDD)

Tupolev Tu-110

In July 1957, less than a year after the Tu-104 entered service, a new transport powered by four turbojets was displayed at Vnukovo Airport, Moscow. This was the Tu-110 which it was stated was to be used on Aeroflot's international services and on certain domestic routes.

The Tu-110 was almost exactly the same in appearance as the Tu-104 but with four engines, lengthened fuselage and increased span. The new aircraft had been designed to have better economics than the Tu-104, increased range and greater safety.

The fuselage was 1·21 m (3 ft 11½ in) longer than that of the Tu-104 and in fact became the prototype of the Tu-104B. In the Tu-104 the engines had been carried by strengthened fuselage frames, but in the

The Tu-110 at Vnukovo Airport, Moscow.

Tu-110 the engines were attached to the wing centre section with consequent saving in fuselage structure weight. The outer wings were the same as those of the Tu-104 but the flaps had increased chord. The addition of the engine-carrying centre section increased the overall span by 2·4 m (7 ft 10½ in).

The four turbojets were designed by A. M. Lyulka's bureau and are believed each to have been of approximately 5,000 kg (11,023 lb) static thrust.

Although the general style of the Tu-110's interior resembled the Tu-104 and Tu-104A, considerable changes were made. The flight deck appears to have been identical to that of the Tu-104 and retained the pressure bulkhead at frame 11. On the starboard side aft of the flight deck was the galley and then extending from frame 15 to the front spar was a cabin with 30 seats. The over-spar galley of the Tu-104 and Tu-104A was replaced by a 15-seat cabin and aft of this was the main cabin with 55 seats. All seats were in triple units on the starboard side and double units to port. The front row of seats in the forward and aft cabins faced aft and there were tables between them and the second seat row. The aft-situated wash rooms and lavatories were the same as those in the Tu-104A.

Drawings of the Tu-110 showed the midship cabin with three windows in each side as in the Tu-104B, but the aircraft seen at Vnukovo in July 1957 had the three roof windows as in the buffet area of the Tu-104 and Tu-104A. The window arrangement in the Tu-110's forward cabin was also identical to that adopted for the Tu-104B.

The cabins of the Tu-110 were pressurized to a differential of 0·5–0·57 kg/sq cm (7·1–8·1 lb/sq in), and a temperature of plus 20 deg C. could be maintained with outside temperatures of minus 50–60 deg C.

To increase baggage and cargo volume, the passenger cabin floor was raised by 14 cm (5½ in) and the underfloor hold floors were lowered by 21 cm (8¼ in). The two holds then had a total volume of 28 cu m (988·8 cu ft), 10 cu m (353·15 cu ft) forward and 18 cu m (635·66 cu ft) aft. The hold doors were on the starboard side.

The interior layout of the Tu-110, including the cargo and baggage holds, was adopted for the Tu-104B.

Removal of the galley from its midship position in the Tu-104 and Tu-104A was of particular advantage because galley stores no longer had to be taken through the passenger cabins.

The main cabin of the Tu-110, looking aft.

It was claimed that the Tu-110 possessed better handling characteristics than the Tu-104 and Tu-104A, that take-off and landing was improved by the larger flaps and that with a 12,000 kg (26,455 lb) payload its tonne-km costs were approaching those of airscrew-turbine powered aircraft.

Payload of the Tu-110 was 12,000 kg (26,455 lb) comprising 7,500–8,000 kg (16,534–17,637 lb) passengers, 1,600–2,000 kg (3,527–4,409 lb) baggage and 2,000–2,900 kg (4,409–6,393 lb) cargo and mail. A 78-seat version was mentioned when the Tu-110 first appeared but no details of the intended layout are known.

The Tu-110 bore a paint scheme identical to that introduced at the same time on the Tu-104A but without the white fuselage top which appeared later. The military red star appeared on the fin in place of the red flag, Aeroflot's name was not carried, and the number 5600 on the fuselage was not prefixed with the civil markings SSSR or the then used Aeroflot letter L.

A small publicity leaflet on the Tu-110 was produced for the Soviet section of the Brussels International Exhibition in 1958, but since then nothing has been heard of the aircraft. However, in that September the

Tu-104B made its initial appearance. It had the same payload as the Tu-110 and seems to have been a twin-engine production version of the four-engine prototype.

Span 36·94 m (121 ft 2½ in); length 40·06 m (131 ft 5 in); height to top of tail 11·9 m* (39 ft 0½ in); sweepback at 25 per cent chord 35 deg; track 13·81 m* (45 ft 3¾ in); wheelbase 14·38 m* (47 ft 2 in).

Maximum payload 12,000 kg (26,455 lb); maximum take-off weight about 79,000 kg (174,165 lb).

Maximum speed 1,000 km/h (539·6 kt) (621·37 mph); cruising speed 800–900 km/h (431·68–485·64 kt) (497·1–559·23 mph).

Take-off field length 1,600 m (5,249 ft); rate of climb with one engine failed at take-off, undercarriage and flaps extended, 5–6 m/sec (984–1,181 ft/min); landing field length 1,200 m (3,937 ft).

Service ceiling 10,000–12,000 m (32,808–39,370 ft).

Maximum payload range at 10,000–12,000 m (32,808–39,370 ft) and 800 km/h (431·68 kt) (497·1 mph) with 1 hr fuel reserve 3,100–3,300 km (1,672–1,780 n.miles), and at 900 km/h (485·64 kt) (559·23 mph) 2,700–2,800 km (1,456–1,510 n.miles); maximum fuel range 5,000 km (2,698 n.miles).

* Unconfirmed.

Tupolev Tu-114

The Soviet Union is undoubtedly a very big country and its size may to some extent have been responsible for the Russian interest in very big aeroplanes. As far back as 1913 Igor Sikorsky built his amazing four-engine cabin biplane *Il'ya Muromets* which was followed by 73 military aircraft of similar design. In 1931 there was the five-engine ANT-14 transport monoplane, in 1934 the eight-engine ANT-20 *Maxim Gorki* and in 1939 the six-engine ANT-20bis which, as the PS-124, saw limited airline service.

The Tu-114 SSSR-76486 at Milan.

The prototype Tu-114, SSSR-L5611, on test in 1958.

When Aeroflot embarked on its big re-equipment programme in 1953, one of the requirements was for a large long-range aeroplane capable of operating over some Soviet domestic trunk routes and of flying long-distance international services. The result was the Tupolev Tu-114, known for a time as the Rossiya (Russia), which made its first flight on 3 October, 1957.

Like the Tu-104, the Tu-114 owed much to a military predecessor, in this case the Tu-20 long-range bomber. There are two versions of the Tu-114, the commercial passenger-carrying aeroplane and the Tu-114D designed to carry urgent mail and freight and a small number of passengers over very long stages.

The Tu-20 was first seen at Tushino in July 1955. It was a very large sweptback monoplane powered by four airscrew-turbines and capable of near-jet performance. The Tu-114D appears to be a straight adaptation of the bomber with only the military equipment removed, a modified interior layout and the addition of cabin windows. The Tu-114, on the

Another view of the Tu-114 SSSR-76486 at Milan. The retractable twin tailwheels can be seen.

other hand, has a completely new fuselage, apart from the nose, a repositioned wing and lowered tailplane.

When the Tu-114 appeared it was the world's largest and heaviest commercial aeroplane and it will only be surpassed in size and weight when the Boeing 747 enters service in 1969 or 1970. The Tu-114 is also the only propeller-driven transport to have a markedly sweptback wing. Its maximum speed of 870 km/h (469·45 kt) (540·59 mph) makes it faster than any other propeller transport by a considerable margin. Whether production of this unorthodox aeroplane, with all doors and emergency exits more than 5 m (16 ft 5 in) above the ground, was really worthwhile it is difficult to say; it certainly presented ground-handling problems when it first appeared at airports outside the Soviet Union.

Passengers disembarking in Moscow from Aeroflot's Tu-114 SSSR-76471.

In layout the Tu-114 is a low-wing monoplane with four very large airscrew-turbines, single fin and rudder, and fully-retractable nosewheel undercarriage.

The wing is an all-metal three-spar structure built in five sections, having anhedral, an aspect ratio of 8·39, and 35 deg sweepback at 25 per cent chord. There are three boundary layer fences on each wing, and the entire trailing edge is occupied by Fowler-type flaps and hydraulically-powered ailerons. There are spoilers on the upper surface immediately forward of the inboard ends of the ailerons. The leading edge has electro-thermal anti-icing. The wing houses the fuel and normal capacity is 72,980 litres (16,053 Imp. gal).

The fuselage is a circular-section semi-monocoque structure which is pressurized for most of its length. Maximum differential is 0·59 kg/sq cm (8·39 lb/sq in) and, as in the Tu-104, there is a pressure bulkhead separating the crew compartment from the remainder of the aircraft. The passenger cabin area has a volume of 332·2 cu m (11,731·53 cu ft), measures 40·3 m (132 ft 2½ in) in length and has maximum width and height of 3·92 m (12 ft 10½ in) and 2·18 m (7 ft 2 in) respectively.

The tail surfaces all have sweepback, the control surfaces are hydraulically powered and the tailplane has variable incidence. The tailplane has electro-thermal deicing but the fin leading edge is protected by a

Main undercarriage units of the prototype Tu-114. (*Flight International*.)

rubber-boot deicer. All control surfaces, including ailerons, incorporate trim tabs.

The massive four-wheel bogy main undercarriage units have a duplicated electrical retraction system and are raised backwards to be housed in large fairings which protrude well aft of the wing trailing edge. The twin nose-wheels are hydraulically retracted backward into the front fuselage and enclosed by two doors. Main-wheel tyre pressure is 9·5 kg/sq cm (135·12 lb/sq in) and nosewheel tyre pressure 8·5 kg/sq cm (120·9 lb/sq in). There is a retractable tail bumper with twin wheels.

Aft cabin of a Tu-114. (*Aeroplane*.)

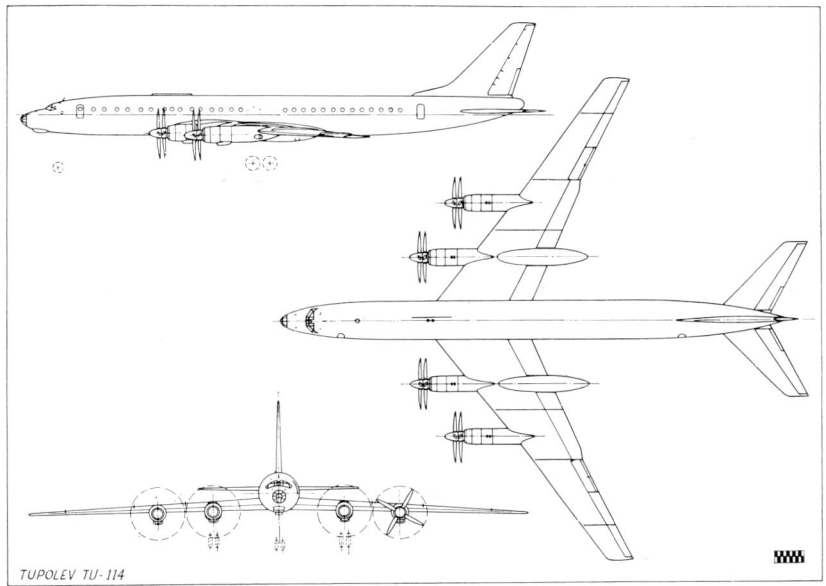

TUPOLEV TU-114

The original engines were four Kuznetsov NK-12M turbines with 14-stage compressor, annular combustion and five-stage axial-flow turbine. These engines each developed 12,000 ehp for take-off and drove 5·6 m (18 ft 4½ in) diameter eight-blade contra-rotating airscrews with large spinners. Later the Tu-114s were fitted with NK-12MV engines which have a nominal power of 12,500 ehp and develop 15,000 ehp for take-off. The airscrews are AV-60N eight-blade contra-rotating automatic-feathering and reverse-pitch type of 5·6 m (18 ft 4½ in) diameter. The blade leading edges have ice protection.

The standard Tu-114 has accommodation for 170 passengers with 146 tourist class seats and 24 first class seats, although the latter can be replaced by 12 sleeping berths; the intercontinental layout is for 120 passengers. In the 170-seat layout there is a forward cabin with 42 rail-mounted seats in seven rows of six-abreast, then comes a cabin with 48 seats arranged in facing groups of six with tables between each block. Between these two forward cabins, and in line with the airscrews, are coat cupboards. Amidship is the galley area, with the kitchen in the lower fuselage and electric lifts to carry food to the upper serving section. Immediately behind the galley is a stairway to the kitchen and a small compartment with two seats, then come two small compartments on each side and these can each have six seats and a folding bunk or two berths. Aft of the small cabins is the main cabin with 54 rail-mounted forward-facing seats in nine rows of six-across. Right aft are lavatories, wash rooms and more coat stowage space. There are also lavatories and cabin crew seats forward of the front cabin.

The 120-seat intercontinental layout has 72 tourist class seats in the forward cabins and 40 first class seats in the rear cabin. The tourist seats are six-abreast and first class five-abreast. There is also an eight-seat saloon. During 1964 Aeroflot was using a 145-seat layout, and there have been reports of a 220-passenger layout for operation over short stages. In this high-density version most of the seats would be in rows of seven- or eight-across with 53 seats forward, 66 amidship, 76 aft and the rest in the galley section and four small cabins.

A 100–104 seat intercontinental version has been reported by Aviaexport and is said to have 60 tourist class seats, 32 first class seats, 8–12 first class sleeping berths and a promenade bar. In this version the front of the forward cabin has been converted to provide two two-seat compartments for crew rest. First class seating is in eight rows of four-across in the aft cabin.

Normal crew comprises two pilots, navigator, engineer, radio operator, two galley staff and at least three stewardesses.

There are front and rear underfloor holds with volumes of 24 cu m (847·55 cu ft) and 46 cu m (1,624·47 cu ft) respectively. The hold hatches measure 80 cm (2 ft 7½ in) by 1·25 m (4 ft 1¼ in) and are from 3·9 m (12 ft 9½ in) to 4·1 m (13 ft 5½ in) above ground level. Passenger entrance doors measure 70 cm (2 ft 3½ in) by 1·3 m (4 ft 3 in), the forward door is 5·7 m (18 ft 8½ in) above ground and the rear door 5·16 m (16 ft 11¼ in). Emergency exits are 55 cm (1 ft 9½ in) by 1·19 m (3 ft 11 in) and from 5·23 m to 5·77 m (17 ft 1¾ in to 18 ft 11¼ in) above ground.

In June 1959 the prototype Tu-114, SSSR-L5611, was shown at the Paris Aero Show and at the end of that month flew nonstop from Moscow to New York in 11 hr 6 min, returning to Moscow in July in 9 hr 48 min. On 10 March, 1960, a Tu-114 made a technical proving flight from Moscow to Khabarovsk. On 17 March, 1961, another proving flight was made over that route while carrying 170 passengers, and on 24 April, 1961, the Tu-114 went into regular service with Aeroflot on the Moscow–Khabarovsk route with a scheduled block time for the 6,800 km (3,669 n.miles) of 8¼ hr.

Aeroflot's Tu-114 SSSR-76464 being readied for departure at Montreal. (*Air Canada*.)

Tu-114 SSSR-76464, with Aeroflot and Japan Air Lines markings, immediately after take-off from Tokyo International Airport. (*Toshihiko Watanabe.*)

On 7 January, 1963, Tu-114s began operating a Moscow–Havana service, with a refuelling stop at Murmansk, and on 25 March, 1963, took over operation of Aeroflot's Moscow–Delhi services. They began operating from Moscow to Conakry and Accra on 19 August, 1965, and on 4 November, 1966, opened a new Aeroflot service between Moscow and Montreal.

In 1966 Japan Air Lines reached agreement with the Soviet Union for operation of a Tokyo–Moscow service and this was undertaken as a joint Aeroflot–Japan Air Lines operation with JAL leasing Aeroflot Tu-114s and using mixed Japanese and Russian crews. A technical proving flight between Moscow and Tokyo took place on 11 August, 1966, and regular weekly services began in April 1967.

Replacement of the Tu-114 began in September 1967 when Ilyushin

Tu-114 SSSR-76464, at Tokyo in April 1967, bearing the markings of Aeroflot and Japan Air Lines. (*Aireview.*)

Il-62s began working one of the twice weekly Moscow–Montreal services. For some time Tu-114s had worked once a week over the Moscow–Paris route and on this and the Moscow–Delhi route they were replaced by Il-62s in October 1967.

Reports that the Tu-114s were in operation on the Moscow–Leningrad and Leningrad–Volgograd services from February 1966 are untrue and are due to a timetable error, Tu-114s having been shown instead of Tu-124s.

The exact number of Tu-114s is not known but appears to be about 30. Production aircraft were registered in the series SSSR-76459 to 76490 and

 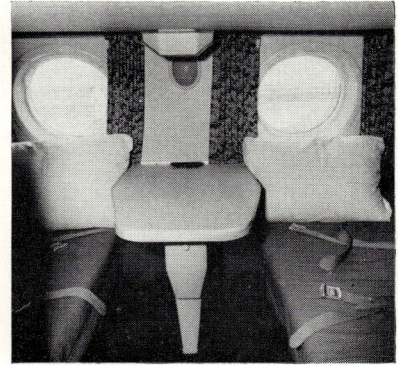

Interior of one of the Tu-114s used on the joint Aeroflot–Japan Air Lines Tokyo–Moscow service. Left, looking across the midship cabin with six-across seating and, right, one of the small cabins with sleeping berths. (*Japan Air Lines.*)

of these 25 are known, one being a Tu-114D. Although the Tu-114 took three and a half years before going into service, and appears to have suffered from vibration and skin panel cracking in its early days, it seems to have operated quite successfully for some seven years. Only one is known to have been lost, having crashed on take-off at Sheremetyevo Airport, Moscow, in very bad weather on 17 February, 1966, on a non-scheduled flight and killing 13 crew and eight passengers.

During the summer of 1967 Tu-114s were flying twice daily in each direction over the Moscow–Khabarovsk route, twice weekly between Moscow and Havana and once weekly over the Moscow–Delhi, Moscow–Paris, Moscow–Tokyo and Moscow–Montreal routes, with some extra flights to Montreal. Without making allowance for extra Montreal flights, a minimum of seven aircraft was required to maintain the services, and weekly scheduled utilization totalled 380 hr 10 min including the time on ground at Murmansk for refuelling on the Havana route. On this basis total annual utilization came to 19,786 hr or 2,826 hr 34 min per aircraft, but this makes no allowance for standby aircraft or time out of service for maintenance.

The Tu-114 set many FAI approved records both before and after its

The Tu-114D SSSR-76462.

entry into service, and details of those still standing in 1967 are shown in Appendix V.

Very little is known about the Tu-114D and only three examples have been recorded, two military aircraft with serials 7801 and 7802 and one civil aircraft SSSR-76462. The fuselage of the Tu-114D is both shorter and of smaller diameter, and all passenger accommodation appears to be aft of the spars which run through the fuselage in a mid-wing position. The tailplane is positioned at the bottom of the fin and is much higher than on the Tu-114. Maximum weight has been reported as 121,920 kg (268,787 lb), the aircraft is known to have accommodation for at least 24 passengers, range is not less than 10,000 km (5,396 n.miles) and long-range cruising speed is about 750–800 km/h (404·7–431·68 kt) (466–497·1 mph).

In the spring of 1958 a Tu-114D made a nonstop flight from Moscow to Irkutsk and back, a distance of 8,500 km (4,586 n.miles) at an average speed of 800 km/h (431·68 kt) (497·1 mph). Also in 1958 a Tu-114D made a flight which covered most of the Soviet Union. The first stage was nonstop from Moscow over Amderma and along the Arctic coast to Poluostrov Taygonos before turning south to Vladivostok. From Vladivostok the aircraft flew nonstop to Tashkent by way of Petropavlovsk-Kamchatskiy. The third stage was from Tashkent to Minsk with overflights of Stalinabad (Dushanbe), Frunze, Alma Ata, Omsk and Dikson on the Arctic coast. The final stage, without an intermediate landing, was from Minsk to Moscow via Leningrad, Tallinn, Riga, Vil'nyus, Kiev, Kishinev, Tbilisi, Yerevan, Baku, Ashkhabad and Rostov.

Another view of the Tu-114D SSSR-76462, showing the slim fuselage and high-set tailplane.

Since these long-distance flights nothing has been heard of the Tu-114Ds and there is no evidence that they ever entered regular commercial service.

Span 51·1 m (167 ft 7¾ in); length 54·1 m (177 ft 6 in); height 15·5 m (50 ft 10¼ in); wing area 311·1 sq m (3,348·64 sq ft); aspect ratio 8·39; sweepback 35 deg at 25 per cent chord; track 13·7 m (44 ft 11½ in); wheelbase 20·69 m (67 ft 10½ in).

Empty weight 91,000–93,000 kg (200,621–205,030 lb); fuel 60,800 kg (134,040 lb); normal payload 15,000 kg (33,069 lb); maximum payload 30,000 kg (66,139 lb); normal take-off weight 164,000 kg (361,558 lb); maximum take-off weight 175,000 kg (385,809 lb); maximum landing weight 135,000 kg (297,624 lb).

Maximum speed at 8,000 m (26,246 ft) 870 km/h (469·45 kt) (540·59 mph); cruising speed at 9,000 m (29,527 ft) 770 km/h (415·49 kt) (478·46 mph); take-off speed at maximum weight 272 km/h (146·77 kt) (169 mph); approach speed 269 km/h (145·15 kt) (167·14 mph); landing speed 205 km/h (110·61 kt) (127·37 mph).

Take-off run at 171,000 kg (376,990 lb) 2,500 m (8,202 ft); take-off run at 175,000 kg (385,809 lb) 2,700 m (8,858 ft); take-off distance to 15 m (49 ft), weight unspecified, 2,850 m (9,350 ft); landing run at 128,000 kg (282,191 lb) 1,400 m (4,593 ft); landing run at 135,000 kg (297,624 lb) 1,550 m (5,085 ft); landing distance from 15 m (49 ft), weight unspecified, using reverse pitch, 1,800 m (5,905 ft).

Service ceiling 12,000 m (39,370 ft).

Range at optimum altitude at 171,000 kg (376,990 lb) take-off weight with 15,000 kg (33,069 lb) payload and 1 hr fuel reserve, 8,950 km (4,829 n.miles); range under same conditions but with 30,000 kg (66,139 lb) payload 6,200 km (3,345 n.miles); maximum fuel range with 13,500 kg (29,762 lb) payload and no fuel reserve, 10,000 km (5,396 n.miles).

Known dates for Tu-114 introductions and technical proving flights:

1959 June 2	Moscow–Tirana (technical proving flight)
1960 March 10	Moscow–Khabarovsk–Moscow (technical proving flight)
1961 March 17	Moscow–Khabarovsk (technical flight with 170 passengers)
1961 April 24	Moscow–Khabarovsk (first regular Tu-114 services)
1962 February 2	Moscow–Vladivostok (technical proving flight)
1963 January 7	Moscow–Murmansk*–Havana
1963 March 25	Moscow–Delhi
1965 August 19	Moscow–Conakry–Accra
1966 August 11	Moscow–Tokyo (technical proving flight)
1966 November 4	Moscow–Montreal
1967 April 17	Moscow–Tokyo (start of joint Aeroflot–Japan Air Lines service, by SSSR-76464)
1967–68 winter	Moscow–Tashkent

* Technical stop.

The Tu-124 SSSR-45006. This view shows how low the aircraft sits.

Tupolev Tu-124

By the end of 1958 Aeroflot had more than two years' operating experience with the Tu-104, the Tu-104A was in service and the 100-passenger Tu-104B was being readied for service in the following spring. Also due to begin operation in 1959 were the large airscrew-turbine powered Antonov An-10 and Ilyushin Il-18, but Aeroflot was still relying on the piston-engined Il-14s for maintenance of many of its services. There was therefore a requirement for a turbine-powered aircraft having about half the capacity of the Tu-104B and possessing much better airfield performance.

To meet this requirement the design bureau under A. N. Tupolev produced plans for the Tu-124 on which actual production began late in 1958 or early in 1959. The prototype, SSSR-45000, made its first flight in June 1960.

The Tu-124 very closely resembled the Tu-104 and was approximately a three-quarter scale version of the older type. The Tu-124 sat very low on the ground, had an untapered trailing edge to the inner section of the wings, and was powered by two turbofans, thus making it the world's first short-to-medium range small jet transport to be powered by turbofans. In fact when Aeroflot introduced the Tu-124 into service between Moscow and Tallinn on 2 October, 1962, it was more than $2\frac{1}{2}$ years ahead of the first western short-haul turbofan-powered transport.

On 10 November, 1962, Tu-124s began operating between Moscow and Ul'yanovsk, on 1 December they replaced Il-14s and Li-2s on Moscow–Gor'kiy services and on 25 December began operating Moscow–Vil'nyus (Vilna) services. Their introduction on the Moscow–Vil'nyus route completed Aeroflot's plan for operation of turbine-powered aircraft linking Moscow with the capitals of all the Soviet Republics.

Aeroflot's Tu-124 SSSR-45013.

The original seating capacity of the Tu-124 was 44 but some time after its initial introduction the Tu-124V appeared as the standard version with 56 seats. The date of introduction of the Tu-124V is not known and therefore it is not possible to distinguish between the two versions in the early years of their service.

During 1963 Tu-124s are known to have been introduced on the following routes: Moscow–Mineral'nyye Vody (1 January), Moscow–Volgograd (21 January), Moscow–Murmansk (13 May), Kuybyshev–Mineral'nyye Vody and Kuybyshev–Sverdlovsk (1 June) and Moscow–Kazan' (16 August). Tu-124s were introduced on the Kuybyshev–Krasnodar route on 21 January, 1964, on the Moscow–Warsaw route on 1 April and Moscow–Adler/Sochi and Moscow–Minsk routes on 15 November that year, and it is reported that the type was serving 24 points in the USSR during the summer of 1964 and operating over a route network of 36,570 km (22,723 miles).

Tu-124s were introduced between Moscow and Belgrade in February 1965 and in the summer of that year were in service with six Aeroflot Directorates and Groups on 44 Federal routes. In 1966 these aircraft were working on more than 70 routes and during the year carried 2 mn passengers.

In the summer of 1967 the Tu-124 was in service in four configurations. The original 44-seat version was being used by Aeroflot's Estonian,

The Tu-124 SSSR-45072 at Le Bourget Airport, Paris, in 1965. Markings are mainly bright blue, with lower fin and fuselage lines red and registration black. (*Flight International.*)

Lithuanian, Moscow Transport and Volga Directorates; the 56-seat Tu-124V was being operated by the Estonian, Georgian, International, Lithuanian, North, Northern Caucasia and White Russian Directorates; Moscow Transport Directorate was using Tu-124Vs with 51 seats on the Moscow–Adler/Sochi route, having reduced the capacity from 56; and 235 Division employed a 38-seat version which was probably the Tu-124K which had earlier been publicized as a 36–37 passenger aircraft. In addition to the Federal services operated by all these Directorates, the North Directorate was using Tu-124s on local services but the seating configuration of these is unknown.

Aeroflot's Tu-124 SSSR-45091 at Leonardo da Vinci Airport, Rome, on 15 April, 1966. (*John Stroud.*)

There have been reports of two mixed-configuration versions of the Tu-124, one carrying cargo forward and having accommodation for 36 passengers aft and the other with increased cargo space and seating for 22 passengers. It is possible that these versions are used on local services but no evidence of their use or even existence has been found.

The Aeroflot Tu-124s, that are known, all bear registrations in the SSSR-45000 series. The following have been identified: 45000, 45003–7, 45013–16, 45018, 45021–3, 45025–6, 45031, 45033, 45038, 45040, 45042–3, 45047, 45052, 45054–6, 45059, 45063, 45067, 45069, 45072, 45082–3 and 45089–92. In addition ČSA purchased a fleet of three Tu-124Vs, OK-TEA *Mělník*, OK-TEB *Centrotex* and OK-UEC *Mladá Boleslav*. The Czechoslovak aircraft went into service on 26 November, 1964, when OK-TEB operated the Prague–Frankfurt-am-Main service. Interflug is reported to have taken delivery of the Tu-124s DM-SDA and DM-SDB, but the airline stated that they would not go into regular service. Two or three Tu-124s have been acquired by the Indian Air Force for VIP transport.

One of the main requirements for the Tu-124 was safe operation from short unpaved runways, and before going into regular service the aircraft was extensively tested in a wide range of climates and while operating from a variety of airfields including those with grass surface.

To meet the airfield requirements the Tu-124 design incorporated a high-lift aerofoil section combined with double-slotted flaps and lift dumpers, under-fuselage wide-span airbrake, undercarriage retraction in 6 sec to meet the engine-out case, low-pressure tyres, and braking parachute in case of an abandoned take-off.

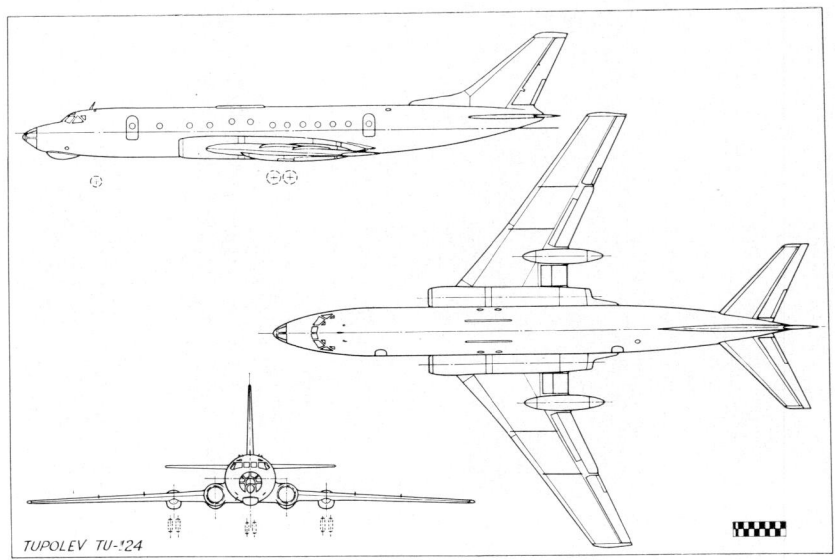

TUPOLEV TU-124

Careful attention in the design was given to servicing for quick turn-rounds, and the aircraft sits very low on the ground thus providing easy access to the wing, engines, undercarriage and electrical systems. Two-point underwing pressure refuelling can be completed in 8 min.

The two-spar wing has slight anhedral, and sweepback at 25 per cent chord is 35 deg. Leading edge sweep is more marked on the inboard sections, and the trailing edge between the engines and undercarriage housings is unswept. The ailerons are manually operated and incorporate spring tabs. The area-increasing and camber-changing double-slotted flaps are lowered 20 deg for take-off and 30 deg inboard and 27 deg 30 min outboard for landing. When retracted the flap slots are protected by a guard which is raised to expose the slots when flaps are lowered. Total flap area is 17·42 sq m (187·5 sq ft). Spoilers on the upper surface of the wing immediately ahead of the flaps open automatically to 52 deg on touch-down. A 3·92 m (12 ft 10½ in) span under-fuselage airbrake of 5·46 sq m (58·77 sq ft) area can be used during descent and for shortening the landing run. This airbrake is deflected 40 deg for landing. The wing leading edge is provided with thermal anti-icing, heat being supplied from the engine compressors. Venting of hot air is via the wing tips. Normal fuel capacity is 13,120 litres (2,886 Imp. gal) housed in 16 tanks, under-fuselage and inner wing tanks being of the flexible rubber-bag type and the outboard tanks of the integral type. The fuel system has been kept as simple as possible, and in case of failure of electric fuel pump motors the pumps are driven by emergency ram-air turbines extended from the wing.

The circular-section fuselage is, with the exception of the tail cone, pressurized to a differential of 0·5–0·57 kg/sq cm (7·1–8·1 lb/sq in) maintaining sea level conditions up to an altitude of 5,000 m (16,404 ft) and 2,400–2,600 m (7,874–8,530 ft) equivalent at 11,000–12,000 m (36,089–39,370 ft). Full pressurization tests including water tank testing with simulated loads were undertaken, and all doors and hatches open inwards. The outer panels of all windows are 16 mm (0·628 in) thick and designed to withstand internal pressures of 7–8 kg/sq cm (99·56–113·79 lb/sq in). The volume of the pressurized area of the fuselage is 122 cu m (4,308·39 cu ft). The cabin length excluding flight deck is 20·7 m (67 ft 11 in), height is 1·95 m (6 ft 4¾ in) and width 2·7 m (8 ft 10¼ in). Baggage and cargo holds are on the main deck, forward on the starboard side and aft between the passenger area and the tail cone. The forward hold has a volume of 6 cu m (211·89 cu ft) and the aft hold is of 8 cu m (282·52 cu ft). Passenger doors are 1·3 m (4 ft 3 in) high and 70 cm (2 ft 3½ in) wide and the loading hatches, on the starboard side, measure 1·2 m by 90 cm (3 ft 11¼ in by 2 ft 11½ in).

The passenger area is divided into three cabins by semi-bulkheads. In the 44-seat Tu-124 the main aft cabin has 24 seats, the midship cabin eight and the forward cabin 12. All seats are in pairs, the front row in each cabin faces aft and there are tables between the two front rows. In the 56-seat Tu-124V there are 32 all forward-facing seats in the aft cabin, 12 forward-facing seats amidship and 12 seats in the front cabin arranged as

Another view of the Tu-124 SSSR-45072 at the 1965 Paris Aero Show. (*John Stroud.*)

ČSA's Tu-124V OK-TEB *Mělník*.

in the 44-seat aircraft. Seats in the Tu-124V are at 750 mm (29·5 in) pitch, are 500 mm (19·68 in) between arm-rest centres, have tip-up seats, reclining backs with folding tables attached and are stressed for 9g. Luggage racks fold against the ceiling and lighting is in two parallel rows. Three standard interior colour schemes are cream, light green and grey. Chair upholstery colours are varied to break up the tunnel effect of the cabin.

In both versions there is a pantry on the port side forward of the front cabin and aft there is a coat cupboard and a lavatory. Cargo capacity was originally 3,291 kg (7,255 lb) but in 1965 this was increased to a maximum of 3,775 kg (8,322 lb).

The cabin is ventilated and heated by air tapped from the engine compressors. Heated air is mixed with cabin air in three underfloor injectors and then enters the cabin through inlets at the base of the walls. Ventilation air is passed through an air cooling turbine and air-to-air heat exchanger and after being cooled to 8–10 deg C. (46·4–50 deg F.) enters the cabin via ceiling ducts.

The 32-seat aft cabin of a ČSA Tu-124V.

The flight deck and navigator's compartment are generally similar to those of the Tu-104 but the crew comprises two pilots and a navigator. A jump-seat is provided for a fourth crew member if required on international flights. Experiments have been made with the Tu-124 operated by a two-man crew.

Flight deck of a ČSA Tu-124V. (*ČSA*.)

The tail unit comprises sweptback tailplane and elevators and fin and rudder with manually-operated control surfaces. Tailplane and fin leading edges have deicing consisting of electrical heating elements encased in fibre-glass between the inner and outer skins.

The main undercarriage units are four-wheel bogies which are hydraulically-retracted backward into streamlined fairings; and the twin nosewheels are also backward retracting. Tyre pressure at 6·5 kg/sq cm (92·45 lb/sq in) is 30 per cent lower than that on the Tu-104, disc brakes are fitted to the main wheels. The nosewheels are steerable and can be turned up to 35 deg each way for taxi-ing and up to 5 deg during take-off and landing. Deflector plates can be fitted behind the nosewheels to prevent mud and gravel being flung into the engine intakes.

The Tu-124 is powered by two 5,400 kg (11,905 lb) thrust Soloviev D-20P two-spool turbofans. The engines are mounted close alongside the fuselage, have heated intakes for ice protection but are not fitted with noise suppressors or thrust reversers. The lower cowlings are hinged to provide access and the engines are so low that they can be reached by a man squatting on the ground. Fuel consumption at 10,000 m (32,808 ft) and 800 km/h (431·68 kt) (497·1 mph) is 1,200 kg/eng/hr (2,645 lb/eng/hr).

The Tu-124 represents an important part of the Aeroflot fleet, appears to be a well thought-out design and has already proved its ditching qualities—SSSR-45021 having made a safe emergency alighting on the Neva in Leningrad after suffering engine failure. Nobody was hurt and the aircraft was towed ashore.

In 1965 the export price quoted for the Tu-124 was $1·45 mn. Operating costs have been quoted by the USSR as just over 22 US cents a ton-mile for the Tu-124V over a stage of 1,287 km (694 n.miles) with 1 hr fuel reserve, engine service life of 1,100 hr and calm air. For the 44-seat Tu-124, ton-mile costs have been given as just over 26 US cents for a stage of about 1,528 km (824 n.miles) under the same conditions.

The number of Tu-124s built is not known but is likely to have been about 100. There has been passenger criticism of the high noise level in the midship and aft cabins, and it has been reported that production of the type was phased out in favour of the Tu-134.

Span 25·55 m (83 ft 10 in); length 30·58 m (100 ft 4 in); height to top of fin 8·08 m (26 ft 6 in); wing area 119·37 sq m (1,284·88 sq ft); aspect ratio 5·5; sweepback at 25 per cent chord 35 deg; track 9·05 m (29 ft 8¼ in); wheelbase 10·55 m (34 ft 7¼ in).

Empty weight 22,900 kg (50,486 lb); maximum fuel 10,500 kg (23,148 lb); maximum payload 6,000 kg (13,227 lb); normal take-off weight 36,500 kg (80,468 lb); maximum take-off weight, on concrete, 37,500 kg (82,673 lb); normal landing weight 32,000 kg (70,548 lb); maximum landing weight 35,000 kg (77,162 lb).

Maximum speed at 8,000 m (26,246 ft) 970 km/h (523·41 kt) (602·73 mph); cruising speed 800–870 km/h (431·68–469·45 kt) (497·1–540·59 mph); landing speed 200 km/h (107·92 kt) (124·27 mph).

Take-off run at 36,000 kg (79,366 lb) at sea level ISA 1,030 m (3,379 ft), at sea level ISA plus 15 deg C. 1,300 m (4,265 ft); take-off distance to 25 m (82 ft) at 36,000 kg (79,366 lb) at sea level ISA 2,100 m (6,889 ft), at sea level ISA plus 15 deg C. 2,800 m (9,186 ft); take-off runway length required at 36,000 kg (79,366 lb) at sea level ISA 1,500 m (4,921 ft), at sea level ISA plus 15 deg C. 1,950 m (6,397 ft); landing run at 32,000 kg (70,548 lb) 930 m (3,051 ft); landing distance from 25 m (82 ft) at 31,000 kg (68,343 lb) 1,100 m (3,608 ft).

Service ceiling 11,700 m (38,385 ft).

Maximum payload range at 10,000 m (32,808 ft) and 800 km/h (431·68 kt) (497·1 mph) against 50 km/h (26·9 kt) headwind with 1 hr fuel reserve 1,250 km (674 n.miles), with 5,000 kg (11,023 lb) payload 1,600 km (863 n.miles); maximum fuel range with 3,000 kg (6,613 lb) payload and 1 hr fuel reserve, 2,100 km (1,133 n.miles).

Although of bad quality this photograph is of interest because it is one of the few showing an Indian Air Force Tu-124—possibly a Tu-124K. (*Indian Air Force.*)

Known dates for Tu-124 introductions and technical proving flights:

Aeroflot

1962 January 11	Moscow–Adler (technical proving flight)
1962 October 2	Moscow–Tallinn (first Tu-124 service, believed by SSSR-45022)
1962 November 10	Moscow–Ul'yanovsk
1962 December 1	Moscow–Gor'kiy
1962 December 25	Moscow–Vil'nyus
1963 January 1	Moscow–Mineral'nyye Vody, Moscow–Stavropol' and Rostov–Simferopol
1963 January 21	Moscow–Volgograd
1963 May 13	Moscow–Murmansk
1963 June 1	Kuybyshev–Mineral'nyye Vody and Kuybyshev–Sverdlovsk
1963 August 16	Moscow–Kazan'
1963 November 2	Moscow–Helsinki and Moscow–Stockholm
1964 January 21	Kuybyshev–Krasnodar
1964 April 1	Moscow–Warsaw
1964 November 15	Moscow–Adler/Sochi and Moscow–Minsk
1965 February	Moscow–Belgrade
1965 May 21	Moscow–Petrozavodsk

ČSA

1964 November 26   Prague–Frankfurt-am-Main   (first ČSA Tu-124 service, by OK-TEB)

Tupolev Tu-134

Some time after the appearance of the Tu-124 there were reports that the Soviet Union was developing the type as the Tu-124A with rear-mounted engines. Then in September 1964 photographs and some details of the new aircraft were first published in the Soviet press and it was known to have been redesignated Tu-134. The press was carried on the Tu-134's 100th test flight which was made on 29 September that year, but it is not known exactly when the type first flew although the likely date is late 1963.

The Tu-134 was designed to provide economic operation over stages of 600–3,200 km (323–1,726 n.miles), to provide standards of passenger comfort comparable to that of the intercontinental jet transports, have a high standard of reliability and be capable of operating from short runways.

Essentially the Tu-134 is an improved Tu-124 with increased capacity, rear-mounted engines and high T-tail. The fuselage diameter of the two types is identical but the Tu-134 fuselage is 1·6 m (5 ft 3 in) longer. The

The second prototype Tu-134, SSSR-45076, landing at Le Bourget during the 1965 Paris Aero Show. The double-slotted flaps and under-fuselage airbrake can be seen. (*R. A. Cole.*)

outer wings of both types appear to be the same except for the ailerons, but the Tu-134 has a lengthened centre section which increases the span by 3·45 m (11 ft 3¾ in). The undercarriage track is in consequence greater, but only by some 40 cm (1 ft 3¾ in).

The two prototypes of the Tu-134, SSSR-45075 and 45076, appear to have been taken straight from the Tu-124 production line and modified, since the Tu-124 registrations run from SSSR-45000; 45072 and 45082 are known to be Tu-124s, while a later Tu-134, shown at Paris in June 1965, was SSSR-65600. This latter Tu-134 may have been the third prototype or the first of 15 pre-production aircraft which were due to begin leaving the factory at about that time. Other Tu-134s which have been identified include SSSR-65601, 65603 and 65610, the last mentioned having been shown in Paris in May and June 1967.

The Tu-134 is known to have undergone very extensive flight testing but the reason for its delay in entering service is not known. The author was told in June 1967 that the Tu-134 would enter service in 2½–3 months' time. No precise date for its entry into service is known but the type first

Tu-134 SSSR-65610 taking part in the 1967 Paris Aero Show. (*R. A. Cole.*)

appeared in Aeroflot's schedules for the winter 1967–68 and valid from 1 November. The routes that the Tu-134 was due to appear on were Moscow–Belgrade, Moscow–Helsinki, Moscow–Kiev–Vienna, Moscow–Stockholm and Moscow–Warsaw. In most cases the Tu-134 was replacing the Tu-124. It is now known that Tu-134s began operating Moscow–Stockholm services in September. Based on past experience it is almost certain that the Tu-134 operated some domestic cargo services before being cleared for passenger operation, but Aeroflot's winter schedules for USSR Federal routes from November 1967 did not show any Tu-134 operations.

Aeroflot's Tu-134 SSSR-65600.

It is likely that Tu-134s have replaced Tu-124s on the production line, and in addition to those being produced for Aeroflot a small number has been ordered by other airlines. ČSA is reported to have ordered 12 for delivery in 1968–70, Interflug three, Malév five and Polskie Linie Lotnicze eight.

Design responsibility for the Tu-134 has been credited to Leonid Selyakov, described as chief designer of the Tupolev design bureau. Deputy designer was Alexandre Arkhangelsky.

In order to provide good take-off and landing performance, the Tu-134, like the Tu-124, is equipped with double-slotted flaps on outer wings and centre section, but repositioning of the engines has allowed an increase in the span of the centre-section flaps. Lift dumpers are fitted to the upper surface of the wing immediately forward of the outboard flaps. The under-fuselage airbrake appears the same as on the Tu-124. Flap setting is 20 deg for take-off and 38 deg for landing. More powerful wheel brakes appear to have replaced the braking parachute fitted to the Tu-124.

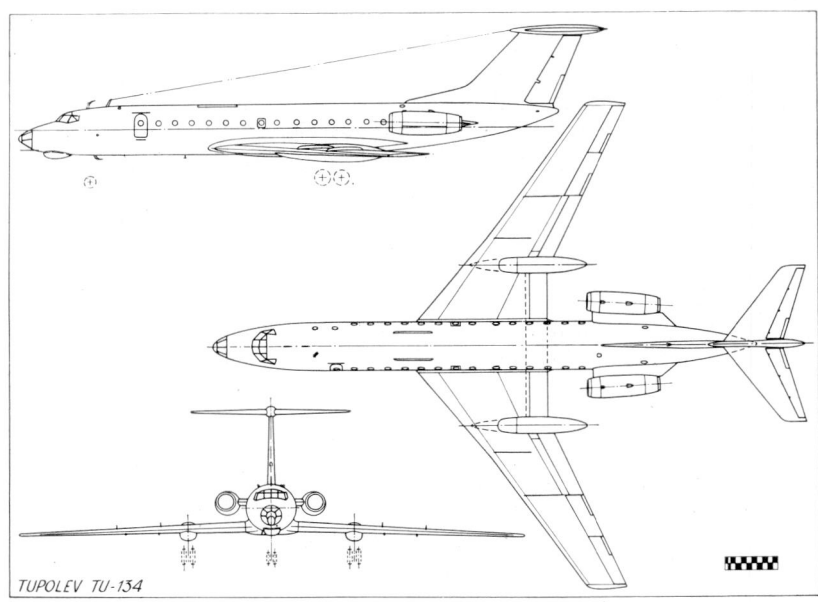
TUPOLEV TU-134

The two-spar wing has 35 deg sweepback at 25 per cent chord. The centre section has an unswept trailing edge but the leading edge sweep is slightly greater than on the outer sections. There is slight anhedral. Chemical milling has been used in the wing structure. The ailerons are divided into two sections on each side and incorporate trim tabs. Nothing is known of the fuel tank layout.

Port main undercarriage unit of the Tu-134 SSSR-45076. (*Flight International*.)

Facing seats at the front of the forward cabin of a Tu-134.

The fuselage, apart from the area aft of the cabin, appears to be identical to that of the Tu-124. Production aircraft have 14 cabin windows on each side. The cabins are pressurized to a differential of 0·57 kg/sq cm (8·1 lb/sq in), giving sea level equivalent up to 6,000 m (19,685 ft) and 2,400 m (7,874 ft) at 12,000 m (39,370 ft).

The single fin and rudder has modest sweepback and carries the markedly swept tailplane and narrow-chord elevators. The tailplane has variable incidence.

Electro-thermal deicing is provided for wing and tail surfaces.

The undercarriage comprises two four-wheel bogies, which retract aft into streamlined fairings, and aft-retracting steerable twin nosewheels. For rough field operation the undercarriage has special 'soft' features which allow the main wheels certain upward and backward movement. Anti-skid disc brakes are fitted to the main wheels. Main-wheel tyre pressure is 6·5 kg/sq cm (92·45 lb/sq in) and nosewheel tyre pressure 5·5 kg/sq cm (78·23 lb/sq in).

The engines are mounted on short horizontal pylons aft of the passenger cabins and are two 6,800 kg (14,991 lb) thrust Soloviev D-30 turbofans with two-stage compressor and three-stage turbine; they are not equipped with thrust reversers or noise suppressors.

Cabin dimensions are: length of seat area 13·85 m (45 ft 5¼ in), maximum width 2·71 m (8 ft 10¾ in) and maximum height 1·96 m (6 ft 5 in). Total volume excluding flight deck is 86·1 cu m (3,040·59 cu ft).

Standard accommodation provides for 72 tourist class passengers in two cabins. All seats are in pairs on either side of the central aisle and all

Tu-134 SSSR-65610 at Vienna Airport. The aircraft is seen in the standard livery applied to production Tu-134s. (*Fritz Kern*.)

except the front row in the forward cabin face forward. Seven rows of seats provide for 28 passengers in the aft cabin, and the 44 seats in the front cabin are in 11 rows. At the rear of the cabin are two lavatories, and right forward on the starboard side is the galley. In the rear fuselage, between the engines, is a cargo hold of 8·5 cu m (300 cu ft) capacity. On the port side forward, opposite the galley, is a hold of 4·5 cu m (158·9 cu ft) capacity. Hold doors measure 1·1 m by 90 cm (3 ft 7¼ in by 2 ft 11·43 in) forward, and 1·2 m by 90 cm (3 ft 11¼ in by 2 ft 11·43 in) aft.

An alternative layout has accommodation for 16 first class and 48 tourist class passengers. In this configuration there are 28 forward-facing tourist class seats in the rear cabin and 20 in a midship cabin. The forward cabin has 16 first class seats with the eight front seats arranged in facing pairs with intervening tables. Tourist class seats appear to be of the same design in both the Tu-124 and Tu-134. Normal operating crew is two pilots and a flight engineer, but the Tu-134 retains the forward navigation compartment of the earlier Tupolev turbine-powered transports and a navigator can be carried as well as a radio operator.

The forward entrance door measures 1·3 m by 70 cm (4 ft 3 in by 2 ft 3½ in) and emergency exits are 60 cm by 60 cm (1 ft 11½ in by 1 ft 11½ in).

The Tu-134 SSSR-65603 taking-off.

The Tu-134 incorporates a flight director system, and provision has been made for the installation of equipment to provide for automatic approach down to 30–40 m (98–131 ft).

If required the Tu-134 can be fitted with an auxiliary power unit to provide ground air-conditioning and engine starting without external power. When it is fitted with an APU the aircraft carries the designation Tu-134A.

A Soviet ditching diagram shows that the aircraft floats with the wing just submerged. The forward door sill is 717 mm (2 ft $4\frac{1}{4}$ in) above the waterline and the overwing emergency exit is 951 mm (3 ft 1·4 in) above water.

Direct operating costs, 1960 ATA method, are quoted as just over 1 US cent per seat-km over stages of 1,000–1,700 km (539–917 n.miles) for a 72-seat aircraft with 2,500 hr annual utilization and 2,500 hr engine overhaul time. Airframe and engine life are quoted as 30,000 hr and 7,500 hr respectively.

Tu-134 SSSR-65611 at Warsaw after flying the inaugural Aeroflot Tu-134 service from Moscow on 2 October, 1967. (*Polskie Linie Lotnicze.*)

Span 29 m (95 ft $1\frac{3}{4}$ in); length 34·35 m (112 ft $8\frac{1}{4}$ in); height to top of tailplane bullet fairing 9·02 m (29 ft $7\frac{1}{4}$ in); wing area 127·3 sq m (1,370·24 sq ft); aspect ratio 6·6; sweepback at 25 per cent chord 35 deg; track 9·45 m (31 ft); wheelbase 13·93 m (45 ft $8\frac{1}{2}$ in).

Empty weight 24,100 kg (53,131 lb); equipped empty weight 26,630 kg (58,708 lb); maximum fuel 13,500 kg (29,762 lb); maximum payload 7,700 kg (16,975 lb); normal take-off weight 44,000 kg (97,003 lb); maximum take-off weight 47,000 kg (103,617 lb); normal maximum landing weight 37,000 kg (81,571 lb); maximum emergency landing weight on paved runway 42,000 kg (92,594 lb).

Maximum cruising speed 920 km/h (496·43 kt) (571·66 mph); economic cruising speed 850–880 km/h (458·66–474·85 kt) (528·17–546·81 mph); long-range cruising speed 780 km/h (420·89 kt) (484·67 mph); take-off speed 248 km/h (133·82 kt) (154 mph); minimum approach speed at maximum landing weight 247 km/h (133·28 kt) (153·47 mph).

Take-off run at 44,000 kg (97,003 lb) at sea level ISA 1,000 m (3,280 ft); balanced take-off field length at maximum weight at sea level ISA 1,900 m (6,233 ft), at sea level ISA plus 15 deg C. 2,200 m (7,217 ft), at 1,500 m (4,921 ft) elevation ISA 2,500 m (8,202

Aeroflot's Tu-134 SSSR-65610 at Schwechat Airport, Vienna. (*Ing. Fred Haubner.*)

ft); landing run at 37,000 kg (81,571 lb) 900 m (2,952 ft); landing distance from 15 m (49 ft) at 37,000 kg (81,571 lb) 1,100 m (3,608 ft); landing runway length required 1,600 m (5,249 ft).

Service ceiling 12,500 m (41,010 ft).

Maximum payload range at 11,000 m (36,089 ft) and 850 km/h (458·66 kt) (528·17 mph) with allowance for 50 km/h (26·9 kt) headwind and 1 hr fuel reserve, 2,400 km (1,295 n.miles); maximum fuel range at 11,000 m (36,089 ft) and 780 km/h (420·89 kt) (484·67 mph) with 5,000 kg (11,023 lb) payload and allowance for 50 km/h (26·9 kt) headwind and 1 hr fuel reserve, 3,250 km (1,753 n.miles).

Known dates for Tu-134 introductions

1967 September 12	Moscow–Stockholm
1967 September 14	Moscow–Kiev–Vienna (by SSSR-65606)
1967 September 16	Moscow–Belgrade
1967 October 2	Moscow–Warsaw (by SSSR-65611)
1967 October 19	Moscow–Helsinki (by SSSR-65606)
1967 December 18	Moscow–Zürich (by SSSR-65610)

It was reported in April 1968 that the Tu-134 is being developed as the Tu-164 with wider fuselage and seating for 120 passengers.

Tupolev Tu-144

It was in November 1962 that Britain and France signed an agreement under which the two countries undertook to work together on the design, development and production of the BAC/Sud Concorde supersonic transport and it must have been at about that same time that the Soviet Union decided to design its supersonic transport, the Tupolev Tu-144. But it was not until the Paris Aero Show in 1965 that the USSR revealed any details of its project. At that show the Russians exhibited a large sectional model of the Tu-144 together with the following information: 'cruise speed 2,500 km/h (1,349 kt) (1,553 mph), range 6,500 km (3,507 n.miles), seating capacity 121, field length 1,900 m (6,233 ft) and operating take-off weight 130,000 kg (286,601 lb)'.

The model of the Tu-144 showed that its general design concept was very close to that of the Concorde and that it is a delta-wing aircraft with fixed geometry and low aspect ratio. The Concorde wing combined with long slender fuselage resulted from 10 years of research and wind tunnel testing in Britain and France and the slender delta was chosen because its sharp sweepback keeps the wing clear of the nose shock-wave, while the low aspect ratio gives minimum drag and adequate fuel stowage. The layout is in fact a modified double-delta. The Tupolev wing, in planform, closely follows the Concorde design but has greater inboard sweepback than the Concorde and less sweepback outboard. However, the Concorde wing has complex cambering and a tapered profile which appears to be less marked in the Soviet design.

The Tu-144 fuselage section is more or less circular with increased radius on the underside, whereas the Concorde fuselage section is almost oval. Both aircraft have a certain amount of area rule. The two types have very similar fins and rudders, neither have tailplanes and both derive pitch and roll control from trailing edge elevons.

Model of the Tu-144 supersonic transport, showing the wing planform and closely-grouped NK-144 turbofans. (*Aviaexport*.)

The Tu-144 has all four engines concentrated close to the centreline whereas the Concorde's engines are grouped in pairs well outboard.

The Concorde undercarriage comprises two four-wheel bogies and twin nosewheels, but the Tu-144 has no less than two sets of three tandem pairs of wheels on each main unit so that, including the twin nosewheels, it has a total of 26 wheels. Both types have a drooping nose to improve the crew's view for take-off and landing.

The dimensions of the Tu-144 are not known but it is bigger than the Concorde which has a span of 84 ft (25·6 m) and a length of 193 ft (58·82 m).

In embarking on their supersonic transport project, the Soviet Union, like Britain and France, decided to use conventional materials and thus obtainable cruising speeds were limited to about Mach 2·2 to 2·3 in order

A model of the Tu-144 showing undercarriage extended and nose lowered. (*Aviaexport.*)

to keep within acceptable temperature limits. By accepting a limiting speed of around Mach 2·3 it has been possible to construct the Tu-144 from conventional light alloys, and titanium has only been used for such high temperature areas as the leading edges and, presumably, the front fuselage which can be expected to reach temperatures of as high as 150 deg C. It is understood that honeycomb construction has not been employed and that as far as possible the structure consists of monolithic panels.

Drag reduction in a supersonic aircraft is of the greatest importance and for this reason the wing has been kept as thin as possible. It is for the same reason that the engines have been grouped as a packet installation aft of the thickest part of the wing, thus reducing the maximum cross section and achieving to some extent area rule. Grouping the engines in this way also helps to give a spanwise spread to the increased pressure at supersonic speed and this increases lift and provides a better lift/drag ratio.

Although Tupolev claims that the chosen ogival wing form and the wing profiles to a large extent reduce centre of pressure changes during transition from subsonic to supersonic flight, it is almost certain that in the Tu-144 some form of fuel trimming must be incorporated. Such a trimming system would involve the use of fore and aft trim tanks. For transonic acceleration, fuel from the front trim tank would be transferred to the main tanks and aft trim tank; for emergency deceleration, fuel would be transferred from the aft trim tank to the forward tank; at the end of cruising flight, aft trim tank fuel would be transferred to the main and forward tanks; and after prolonged subsonic flight, forward tank fuel would be transferred to the main tanks before landing. This is the system employed in the Concorde, and the two designs are so alike that it is difficult to believe that the Tu-144's fuel system can be very different.

In designing the fuselage of a supersonic transport there is very little freedom in choice of dimensions. The fuselage has to be very long in relation to its diameter if acceptable drag figures are to be obtained, this means that the supersonic transport must have a long finely-pointed nose. In turn this means that pilot view is extremely poor, and to overcome this problem BAC/Sud with the Concorde, Boeing with the 2707 and Tupolev with the Tu-144 have all had to incorporate a downward drooping nose so that take-offs and landings can be made with normal flight deck windows exposed to provide adequate view. In cruising flight the Tu-144 is said to provide pilot vision through panels in the movable nose and these, in the original design, are known to provide very inadequate view.

The Concorde was to have provided its pilots with the same limited view in cruising flight but pilot resistance has been so strong that the design has been changed. It would seem reasonable to expect that the Tu-144 design, too, will have to be changed.

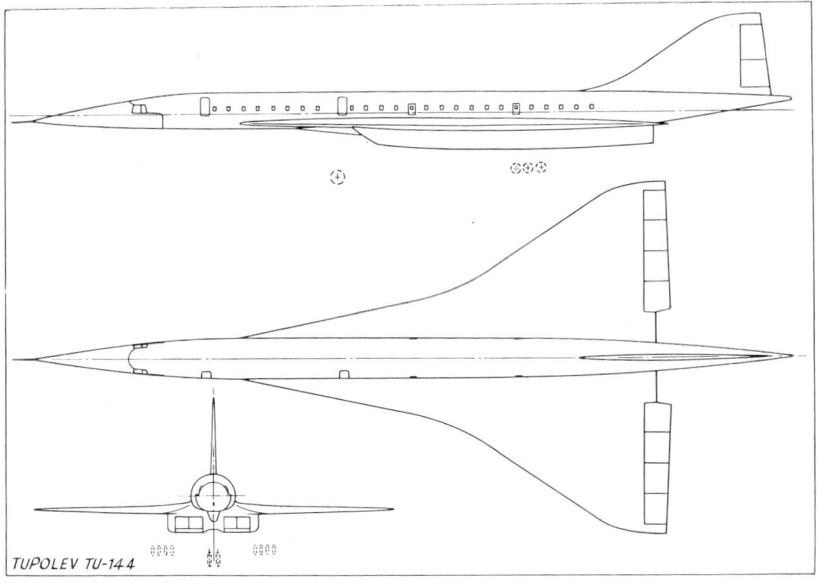

TUPOLEV TU-144

Pilots have also shown concern at the possibility of the nose failing to droop for landing, and in the case of the Tu-144 it is known that three independent systems provide for any failure of nose actuation.

Little is known of the actual structure of the Tu-144, but the wing is known to have four-section elevons on each side, each with its own hydraulic booster, and that leading edge ice protection is provided by hot air bled from the engine compressors, with insulation preventing the heat

Part of the 18-seat first class cabin of the Tu-144 mock-up. (*Aviaexport.*)

being transmitted to the main wing structure. The main undercarriage units retract forward into the wing and the nosewheels retract into a fairing immediately forward of the engine air intakes. The Tu-144's rudder is in two sections and it is likely that one half will be locked when the aircraft is flying at supersonic speed.

The four engines are Kuznetsov NK-144 turbofans developed from the NK-8 used in the Il-62. The NK-144 is a two-spool engine with a bypass ratio of 1:1 and a compression ratio of 15:1. It is reported to have five-stage low-pressure compressor, 11-stage high-pressure compressor, single-stage high-pressure turbine, two-stage low-pressure turbine and annular combustion chamber. Turbine blades are air-cooled. The engine is said to be 5·2 m (17 ft $0\frac{3}{4}$ in) in length and 1·5 m (4 ft 11 in) in diameter. Dry weight is reported as 2,850 kg (6,283 lb). Dry thrust is 13,000 kg (28,660 lb) and take-off thrust with reheat is 17,500 kg (38,581 lb). Tupolev has stated that reheat will only be used to meet the engine-out case, and that the NK-144 is fitted with noise suppressors and thrust reversers. The wing main-spar box houses 70,000 kg (154,324 lb) of fuel.

The engine intakes have variable-geometry, which is essential for supersonic operation, and, when a new model of the Tu-144 was shown at the 1967 Paris Aero Show, it was seen that the engine intakes had been bifurcated to divide the engines into pairs. An aerodynamic strake had also been added to the centreline of the undersurface of the large engine-housing structure. The nosewheel fairing also probably serves as a shock-body.

The flight deck is laid out for three-crew operation, with two pilots and flight engineer. Aft of the flight deck is a considerable area devoted to the housing of avionic equipment. Immediately behind this section is the front entrance vestibule with the door on the port side and just forward of it a lavatory. Baggage and coat-hanging space is provided both on the starboard side and just aft of the door. The forward cabin can provide accommodation for 18 first class seats, three-across with double seats on the starboard side, or 40 tourist class seats, five-across with triple units to starboard. Then comes another entrance door which is over the highly-swept section of the wing and in this area is the galley and more coat-hanging space. The aft cabin has 80 tourist class seats with 12 rows of five-across and at the rear five rows with double seats on each side. At the rear of this cabin are two lavatories. There are eight rectangular windows in each side of the forward cabin and 17 in each side of the main cabin. Two windows in each side of the main cabin are incorporated in overwing emergency exit doors. Inflatable slides are provided at exits and dinghies will be carried for over-water operation.

Aft of the passenger accommodation is the main baggage hold which can take four shaped containers. These are loaded via an under-fuselage hatch, according to an article by A. N. Tupolev, or over the wing by mechanical means, according to Soviet representatives at the Paris Aero Show.

It is understood that three Tu-144 prototypes were being built, and in the autumn of 1967 it was stated that the first flight would be made in the first quarter of 1968. Entry into Aeroflot service is planned for the early 1970s.

Dimensions are not known but cabin height has been reported as 1·95 m (6 ft 4¾ in) forward and 1·86 m (6 ft 1¼ in) aft, and interior maximum width as 3·2 m (10 ft 6 in). Fuel 70,000 kg (154,324 lb); maximum payload 12,000 kg (26,455 lb); take-off weight 130,000 kg (286,601 lb). Cruising speed Mach 2·35–2,500 km/h (1,349 kt) (1,553·4 mph); take-off run 1,900 m (6,233 ft); runway length required 2,700 m (8,858 ft); landing run 1,500 m (4,921 ft); ceiling 20,000 m (65,617 ft); maximum range 6,500 km (3,507 n.miles).

Tupolev Tu-154

It first became known in 1966 that the Tupolev design bureau had been working on plans for an aircraft to replace the numerous An-10s, Il-18s and Tu-104s in service with Aeroflot on trunk routes. This new project was the Tu-154, and it was stated early in 1966 that prototypes and pre-production aircraft were already under construction.

The Tu-154 is the biggest of the Tupolev turbojet-powered transports, has three rear-mounted engines and T-tail and is most nearly comparable to the Boeing 727-200. The Soviet aircraft is somewhat larger, heavier and more powerful than the Boeing, but has less passenger accommodation.

The wing planform and undercarriage housing used in the Tu-154 is similar to that of the Tu-124 and Tu-134, but apart from those features the aircraft bears little resemblance to earlier Tupolev designs.

Design requirements were for the most economic and safe aircraft capable of high speed and having the ability to operate from second-class aerodromes. In consequence the design team set out to produce an aircraft with the range of the Il-18, the speed of the Tu-104 and the take-off and landing performance of the An-10. High-altitude operation was considered necessary to reduce air route congestion, and some limits were put on the dimensions of the resulting Tu-154 because it had to be able to use hangars and maintenance docks provided for the Tu-104Bs.

As in other parts of the world, the three-engine layout was chosen as a compromise between the economy of two engines and the reliability of four engines. In order to achieve good take-off performance from short runways and high and hot airports, a high thrust/weight ratio was chosen—0·35 to 0·36 compared with the more usual 0·22 to 0·27. The wing area was dictated by the need to cruise at 12,000 m (39,370 ft) to avoid interference with traffic at lower levels, and to keep the approach speed as low as possible. Thus to some extent a slight economic penalty has been accepted in order to gain greater operational flexibility and particularly the ability to operate from secondary airports.

The wing is a three-spar structure with 35 deg sweepback at 25 per cent chord. The centre section is relatively thick and without sweep on its trailing edge. Hydraulically-operated triple-slotted flaps extend from the undercarriage fairings to the narrow-span ailerons and between the undercarriage fairings and the fuselage. Electrically-operated slats occupy most of the leading edge. Immediately forward of the flaps are a series of interceptors. On the outer wings are three interceptors and each centre section has one. The ailerons provide normal lateral control at cruising speeds but at low speed the outer interceptors are activated when the ailerons reach an angle of 8 deg. The other two outer interceptors each side act as airbrakes and on landing combine with the centre-section

A model of the Tu-154, showing the engine layout, high T-tail and undercarriage housings. (*Aviaexport*.)

interceptors to serve as airbrakes and lift-dumpers. Wing-root leading edges are protected from ice by hot air bled from the engine compressors, and the slats are protected by an electro-thermal system.

The outer wings contain five integral fuel tanks with a capacity of 33,150 kg (73,083 lb) and there are four centre-section bag tanks with 7,150 kg (15,763 lb) capacity. Fuel is pumped into a collector tank, and six electric pumps supply fuel from the collector tank to the engines. There is a pressure refuelling point under the starboard wing, and the normal fuel load can be delivered from a KZ-22 refueller in 13 min or from two KZ-16s in 27 min.

The fuselage is a circular-section semi-monocoque structure most of which is pressurized to a differential of 0·63 kg/sq cm (8·96 lb/sq in). All fuselage skins are chemically machined and in door and window cut-out areas thickness is 8 mm (0·31 in). Actual cabin dimensions are not known but external fuselage diameter is 3·8 m (12 ft $5\frac{3}{4}$ in), and the cabin is wide enough to allow six-abreast tourist class seating. The unpressurized nose section houses weather-warning radar. The flight deck is a completely new Tupolev design and provides for a normal crew comprising two pilots and a flight engineer, but provision is made for carriage of a navigator for over-ocean or northern routes with poor navigational facilities. Immediately aft of the flight deck is the forward entrance vestibule which contains a bench-type seat for three cabin crew and a crew coat compartment. On the starboard side are a lavatory and a space for passengers' coats. Next comes the forward cabin and this is separated from the aft cabin by the buffet

First photograph to be released of the Tupolev Tu-154. Taken on 24 April, 1968, this picture shows the aircraft during undercarriage retraction tests. In the background is the prototype Tu-144 supersonic transport SSSR-68001. (*Press Association Ltd.*)

area, rear entrance vestibule with three more cabin-crew seats and another passengers' coat cupboard. Aft of the main cabin are three lavatories.

The tail unit comprises a swept fin and rudder above which is mounted the sweptback tailplane and elevators. The tailplane is movable and the elevators are not interconnected, but in case of a failure one elevator combined with tailplane setting is sufficient to maintain control. Fin and tailplane leading edges have hot-air deicing.

All controls are power operated. There is a triplicated hydraulic system but no manual reversion.

The main undercarriage units are six-wheel bogies with three pairs of wheels in tandem. Retraction is backward into typical Tupolev undercarriage fairings. The twin nosewheels are steerable, with up to 55 deg movement from the centreline. All tyres have a pressure of 8 kg/sq cm (113·79 lb/sq in). It is claimed that the wheel loading of the Tu-154 is such that a runway concrete thickness of 18–20 cm (7–7·8 in) is sufficient, and that the aircraft can operate from natural surfaces having a bearing strength of more than 8 kg/sq cm (113·79 lb/sq in).

The three engines are Kuznetsov NK-8-2 turbofans each developing 9,500 kg (20,943 lb) of thrust at take-off. Nominal thrust at 11,000 m (36,089 ft) is 2,650 kg (5,842 lb) with a specific fuel consumption of 0·79 kg/kg (0·79 lb/lb). Two engines are mounted on the fuselage sides immediately aft of the cabin area and are easily accessible for maintenance. The third engine is in the rear fuselage, is accessible through an underside hatch, and receives its air via an intake above the fuselage and a curved duct similar to that in the Boeing 727 and Hawker Siddeley Trident. Engine intakes and the centre engine duct have hot-air deicing. The outer engines are fitted with thrust reversers. Noise level at take-off is said to be 90–95 PNdB. In cruising flight engine power is reduced to 70–75 per cent nominal power with consequent increased life. Engine life is quoted by Tupolev's bureau as 10,000 hr with 2,000 hr time between overhauls, but work is in hand to raise the TBO figure to 5,000 hr. An auxiliary power unit is mounted above the centre engine.

The basic Tu-154 has 158 tourist class seats, with accommodation for 54 passengers in the forward cabin and 104 in the main cabin. All seats are triple units on each side of the centre aisle except for the two back rows in the main cabin which are in pairs. The front row in each cabin is backward facing. A high-density 164-seat version would have 60 passengers in the front cabin. Other configurations so far mentioned are for 128, 134 and 146 passengers. The 128-seat aircraft would have 24 first class seats in the forward cabin and retain the 104-seat tourist class layout in the main cabin. Layout of the 134- and 146-seat versions is not known. Winter seating is reduced by the removal of the eight rear seats from the main cabin to provide space for 80–82 coats.

All seats are installed on rails with locking points spaced at 30 mm (1·18 in). Seat pitch in the 158-seat version is 750 mm (29½ in). When the forward cabin has 24 first class seats installed these are at 1,020 mm (39·4 in) pitch. Table units with stowage space have been provided between each pair of first class seats. The cabin walls consist of fire-resistant synthetic materials, windows are spaced at 500 mm (19·6 in), compared with 1 m (39·3 in) on the earlier Tupolev jet transports, and pull-down shades are fitted.

Except when outside temperatures exceed 35 deg C. (95 deg F.) combined with relative humidity of more than 50 per cent, cabin temperature is

Part of the first class forward cabin of the Tu-154. This photograph was probably taken in the mock-up. (*Aviaexport*.)

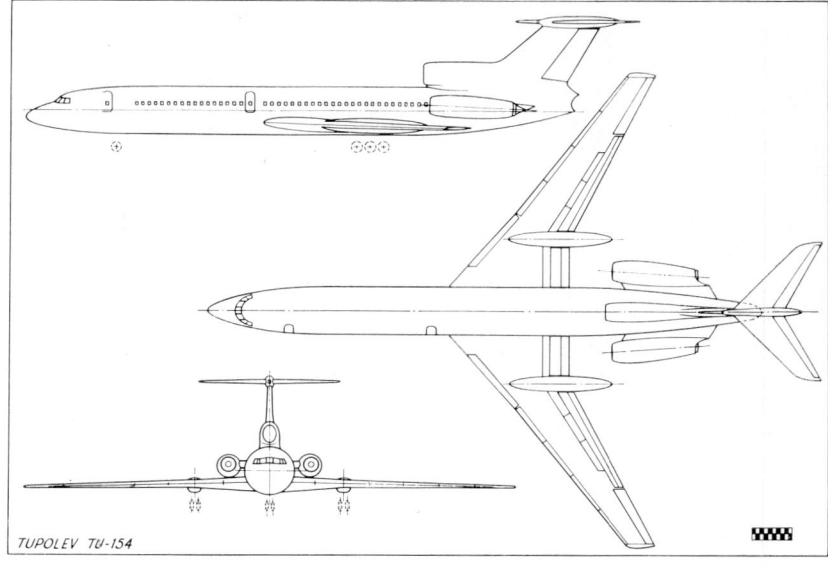

TUPOLEV TU-154

kept at 20 deg C. (68 deg F.). Cabin heating air is passed behind the wall panels, and ventilating air, at 10 deg. C. (50 deg F.), is supplied from individual outlets under the hat racks.

Passenger entrance doors measure 1·75 m (5 ft 9 in) by 80 cm (2 ft $6\frac{1}{2}$ in), and there are two doors on the starboard side which measure 1·28 m (4 ft $2\frac{1}{4}$ in) by 61 cm (2 ft). There are also overwing emergency exits on each side and one in each side of the forward cabin. The doors are provided with inflatable stairs and escape chutes. For over-water operation four dinghies each capable of carrying 30 people are carried in the centre section and in the rear of the undercarriage fairings. If the number on board exceeds 120, additional dinghies are carried in the entrance vestibules. Emergency descents from 12,000 m (39,370 ft) to 5,000 m (16,404 ft) can be made in 2–$2\frac{1}{2}$ min, with wheels lowered and airbrakes extended.

There are two pressurized underfloor holds with a total capacity of 38·5 cu m (1,359·61 cu ft) and one unpressurized hold of 5·6 cu m (197·75 cu ft). Containers can be carried in the pressurized holds.

The flight deck is very fully equipped, and navigational aids include Doppler. Initially the Tu-154 will be cleared for ICAO Category II operation, and continuing development work will lead to complete autoland for Category III operations.* Windscreen panels are electrically

* ICAO Category I. 60 m (200 ft) ceiling and 800 m ($\frac{1}{2}$ mile) RVR (Runway Visual Range)
ICAO Category II. 30 m (100 ft) ceiling and 400 m ($\frac{1}{4}$ mile) RVR
ICAO Category IIIA. 200 m (700 ft) RVR
ICAO Category IIIB. 45 m (150 ft) RVR

heated and can withstand strikes from birds of up to 1·7 kg (3¾ lb) in weight.

Investigation of the Tu-154's stability has been carried out up to angles of attack as high as 45–60 deg. A stick-shaker and a stick-pusher is fitted.

Airframe life is designed for 30,000 hr with 5,000 hr overhaul time. Break-even load factor, at fares in force in 1966, was estimated at 30–35 per cent.

Under study are developments of the Tu-154. One possibility is a freight version with strengthened floor, a large door measuring 2·1 m by 3·4 m (6 ft 10¾ in by 11 ft 1¾ in) and pallet loading. This variant would carry a 25,000 kg (55,116 lb) payload over 2,000–2,500 km (1,079–1,349 n.mile) stages at 900 km/h (485·64 kt) (559·23 mph). Another study is for a stretched-fuselage aircraft with seating for 240–250 passengers.

Span 37·55 m (123 ft 2½ in); length 47·9 m (157 ft 1¼ in); height to top of tail 11·4 m (37 ft 5 in); wing area 201·5 sq m (2,168·9 sq ft); sweepback at 25 per cent chord 35 deg; track 11·5 m (37 ft 8½ in); wheelbase 18·92 m (62 ft 0¾ in).

Structure weight 40,200 kg (88,625 lb); equipped empty weight with crew 41,700 kg (91,933 lb); maximum fuel excluding centre-section tanks 33,150 kg (73,083 lb); maximum fuel including centre-section tanks 40,300 kg (88,846 lb); normal payload 19,000 kg (41,888 lb); maximum payload 21,500 kg (47,399 lb); normal take-off weight 80,000 kg (176,370 lb); maximum take-off weight 86,000 kg (189,598 lb); normal landing weight 62,000–67,500 kg (136,687–148,812 lb); maximum allowable landing weight under certain unspecified conditions 78,000–80,000 kg (171,961–176,370 lb).

Cruising speed at 11,000 m (36,089 ft) Mach 0·94–1,000 km/h (539·6 kt) (621·37 mph); economic cruising speed at 11,000–12,000 m (36,089–39,370 ft) Mach 0·85–900 km/h (485·64 kt) (559·23 mph); long-range cruising speed at 11,000–12,000 m (36,089–39,370 ft) Mach 0·8–850 km/h (458·66 kt) (528·17 mph); approach speed at 62,000 kg (136,687 lb) with flaps down and slats open 220 km/h (118·71 kt) (136·7 mph); approach speed at 67,500 kg (148,812 lb) with flaps down and slats open 230 km/h (124·11 kt) (142·91 mph); touchdown speed at 62,000 kg (136,687 lb) 215 km/h (116 kt) (133·59 mph); touchdown speed at 67,500 kg (148,812 lb) 225 km/h (121·41 kt) (139·8 mph).

Take-off run at 80,000 kg (176,370 lb) at sea level ISA, 800 m (2,624 ft), with 1,250 m (4,101 ft) required runway length; take-off run at 86,000 kg (189,598 lb) at sea level ISA, 950 m (3,116 ft), with 1,450 m (4,757 ft) required runway length; landing run with thrust reverse on side engines 600–700 m (1,968–2,296 ft); landing run without thrust reverse 700–800 m (2,296–2,624 ft).

Range with 80,000 kg (176,370 lb) take-off weight and 19,000 kg (41,888 lb) payload at 900 km/h (485·64 kt) (559·23 mph) with reserves, 2,850 km (1,537 n.miles); range with 86,000 kg (189,598 lb) take-off weight and 19,000 kg (41,888 lb) payload at 900 km/h (485·64 kt) (559·23 mph) with reserves, 4,000 km (2,158 n.miles); maximum fuel range, with full centre-section tanks and 6,000 kg (13,227 lb) payload, 7,000 km (3,777 n.miles).

Data is based on design figures and performance estimates by the Tupolev design bureau, issued before the first flight of the prototype.

Yakovlev Yak-12

The Yak-12 is a small multi-purpose monoplane capable of carrying pilot and three passengers. It is the Soviet equivalent of the British Auster series and its inclusion in a book of transport aircraft may well be questioned, but in fact the Yak-12 is an important element in Soviet air transport and provides a service to large numbers of communities which would otherwise be isolated from the main airline network.

A Yak-12R of Dosaaf. Full-span fixed leading edge slats are fitted.
(*Courtesy William Green.*)

The original Yak-12 made its debut in 1944 and was of rather Aeronca-like appearance. It was a strut-braced high-wing monoplane with single fin and rudder and non-retractable spatted undercarriage. This early Yak-12 was of mixed construction, had a deep fuselage and was powered by a 145/160 hp Shvetsov M-11 five-cylinder air-cooled radial engine with helmet fairings over the cylinder heads. A revised version appeared in 1947 and this had a number of modifications including fixed leading edge slats, shallower fuselage and removal of the spats. Production of this version began in 1948 and it is believed to have been supplied in quantity to the Soviet Air Force for liaison duties.

As a trainer the Yak-12 was found to be underpowered; therefore in 1952 the Yak-12R version was produced, with increased span and area,

The Aeroflot Yak-12M ambulance SSSR-07870 over the mountains of Tadzhik SSR.

lengthened fuselage and the 240 hp Ivchenko AI-14R nine-cylinder air-cooled radial engine with two-blade controllable-pitch metal airscrew. This improved version was supplied to the Soviet Air Force, Aeroflot and other Soviet aviation organizations. At least some of these Yak-12Rs had an under-fuselage hook which acted as a brake to reduce landing run.

The next variant was the Yak-12M with all-metal structure and a fin extension along the top of the fuselage. Although it was in service for some time, the Yak-12M proved to have inadequate payload and a comparatively short range. As a result a new and modified model was designed to meet Aeroflot's requirements, this was the Yak-12A which appeared in 1957.

The Yak-12A had a new wing with taper on the outboard trailing edge, a reduction in area, reinforced leading edge, automatic leading edge slats

Latest version of the Yak-12, the Yak-12A with tapered outer wings and single bracing strut. The Aeroflot example shown is SSSR-72806. (*Courtesy William Green.*)

and single wing-bracing struts. Fuel and oil capacity was increased and a number of changes were made to simplify maintenance and servicing. There were additional cabin windows and the cabin interior was generally improved. According to a Russian description the cabin walls and doors were insulated with hair, presumably Yak hair!

The use of automatic slats combined with the other aerodynamic refinements improved handling and performance, and this performance improvement, together with increased fuel, cut the tonne-km costs of the Yak-12A to 8 roubles 50 kopecks below that of the Yak-12M. The auto-slats come out at 105 km/h (56·65 kt) (65·23 mph) and close at 120

A Soviet Air Force Yak-12A.

km/h (64·75 kt) (74·56 mph) in the climb, and in a glide they open at 120 km/h (64·75 kt) (74·56 mph) and close at 140 km/h (75·54 kt) (86·99 mph). They do not have any effect on the aircraft's stability.

The wing of the later versions was a two-spar metal structure and all but the Yak-12A had wings of parallel chord. Their wings were braced by V struts, had full-span slotted ailerons and flaps and full-span leading edge slats. The Yak-12A wing has trailing edge taper outboard of the slotted flaps and single struts, and the leading edge slats only occupy the outer sections of the wing. The fuselage is a rectangular-section welded steel-tube structure and most aircraft are believed to have had fabric covering, but the Yak-12A may have a ply-covered fuselage. The tail surfaces are of metal construction with fabric covering and the tailplane is wire braced. The undercarriage comprises divided main units and non-retractable tailwheel.

All production aircraft since the original Yak-12s have had the AI-14R engine with controllable intake shutters, and a two-blade V-530-D11 variable-pitch metral airscrew with square tips.

The Yak-12s are used for a wide range of duties including training, forest patrol, glider-towing, passenger and cargo transport, ambulance work and agricultural spraying and dusting. In the passenger role the earlier Yak-12s had four seats arranged in two tandem pairs but in the Yak-12A the rear seat is of the bench type for two passengers. For cargo work the cabin has attachment points for lashing down a load of up to 300 kg (661 lb). As an ambulance the Yak-12 can carry a stretcher, a sitting

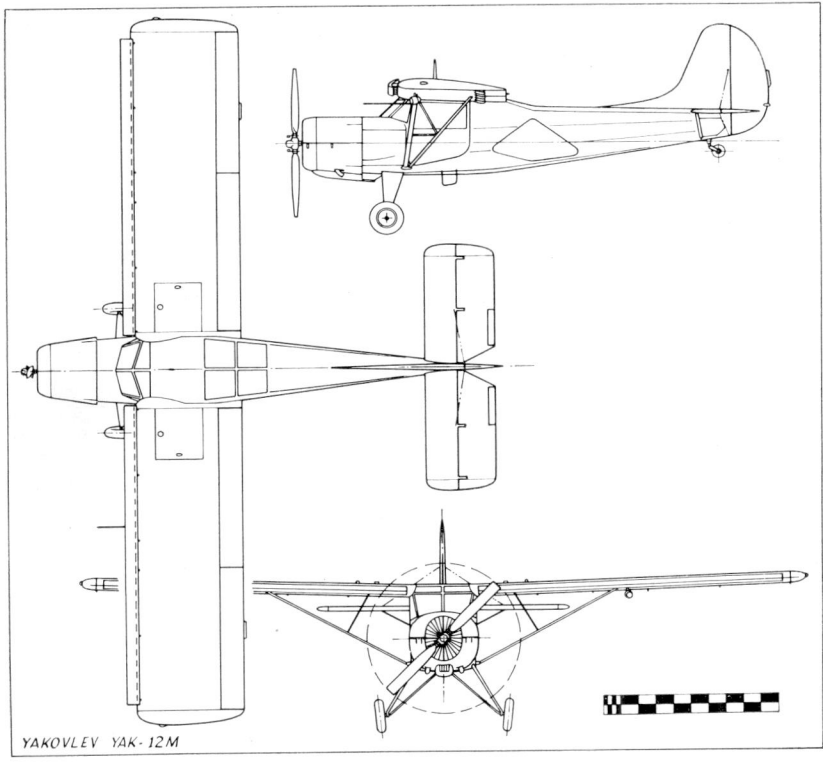

YAKOVLEV YAK-12M

patient and a medical attendant, the stretcher being loaded through a triangular door on the port side aft of the cabin. For agricultural work the Yak-12 has a 470 litre (103 Imp. gal) chemical container and underwing spray-bars or ejector nozzles. Pressure is supplied by a small propeller mounted beneath the fuselage and directly in the slipstream.

Aeroflot has used large numbers of Yak-12s of different types and they are known to have served the Eastern Siberia, Kazakh, Moldavian, Tadzhik, Turkmen, Ukrainian, Uzbek and White Russian Directorates, in many cases on local services. Although they are being replaced, numbers are still in service and the Turkmen Directorate was certainly using them in 1967. Some aircraft are known to have been operated on skis.

Yak-12s have been exported to a number of countries. Soviet production ceased in 1960 but the type was also produced in Poland and developed as the multi-purpose PZL-101 Gawron (Rook) which is powered by the AI-14R, completing its trials in 1958.

Yak-12

Span 12 m (39 ft 4½ in); length 8·45 m (27 ft 8½ in); height 3·12 m (10 ft 2¾ in); wing area 22 sq m (236·8 sq ft). Empty weight 770 kg (1,697 lb); useful load including fuel and pilot 430 kg (948 lb); take-off weight 1,200 kg (2,645 lb). Maximum speed at sea level 200 km/h (107·92 kt) (124·27 mph); maximum cruising speed 161 km/h (86·87 kt) (100·04 mph); landing speed 70 km/h (37·77 kt) (43·49 mph); take-off run 100 m (328 ft); landing run 70 m (229 ft); service ceiling 4,000 m (13,123 ft); maximum payload range 1,000 km (539 n.miles).

Yak-12R

Span 12·6 m (41 ft 4 in); length 9 m (29 ft 6¼ in); height 3·12 m (10 ft 2¾ in); wing area 23·86 sq m (256·82 sq ft). Empty weight and payload not known. Take-off weight 1,285 kg (2,832 lb). Maximum speed at sea level 184 km/h (99·28 kt) (114·33 mph); landing speed 62 km/h (33·45 kt) (38·52 mph); take-off run 75 m (246 ft); landing run 65 m (213 ft); payload range 500 km (269 n.miles).

Yak-12M

Span 12·6 m (41 ft 4 in); length 9 m (29 ft 6¼ in); height 3·12 m (10 ft 2¾ in); wing area 23·86 sq m (256·82 sq ft). Empty weight 1,026 kg (2,262 lb); fuel 138 kg (304 lb); maximum payload 250 kg (551 lb); maximum take-off weight 1,450 kg (3,196 lb). Maximum speed at sea level 180 km/h (97·13 kt) (111·85 mph); maximum cruising speed 140 km/h (75·54 kt) (86·99 mph); economic cruising speed 127 km/h (68·52 kt) (78·91 mph); long-range cruising speed 120 km/h (64·75 kt) (74·56 mph); take-off speed 80 km/h (43·16 kt) (49·71 mph); landing speed 73 km/h (39·39 kt) (45·36 mph); take-off run 153 m (501 ft); landing run 120 m (393 ft); full load rate of climb 4·1 m/sec (807 ft/min); service ceiling 4,160 m (13,648 ft); maximum payload range in still air without reserves 450 km (242 n.miles).

Yak-12A

Span 12·6 m (41 ft 4 in); length 9 m (29 ft 6¼ in); height 2·33 m (7 ft 7¾ in); wing area 22·61 sq m (243·36 sq ft); track 2·2 m (7 ft 2½ in). Empty weight 1,059 kg (2,334 lb); maximum fuel 166 kg (365 lb); maximum payload 300 kg (661 lb); maximum take-off weight 1,588 kg (3,500 lb). Maximum speed at sea level 215 km/h (116 kt) (133·59 mph); maximum cruising speed 170 km/h (91·73 kt) (105·63 mph); economic cruising speed 155 km/h (83·63 kt) (96·3 mph); long-range cruising speed 150 km/h (80·94 kt) (93·21 mph); take-off speed with 20 deg flap 80 km/h (43·16 kt) (49·71 mph); landing speed with 40 deg flap 90 km/h (48·56 kt) (55·92 mph); take-off run with 20 deg flap 153 m (501 ft); landing run on grass with use of brakes 131 m (429 ft); full load rate of climb 3·6 m/sec (708 ft/min); service ceiling 4,000 m (13,123 ft); maximum payload range in still air without reserves 1,070 km (577 n.miles); maximum payload range against 10 km/h (5·4 kt) headwind and with 1 hr fuel reserve, 600–800 km (323–431 n.miles) depending on percentage of power used.

Yakovlev Yak-16

A Soviet transport aircraft about which comparatively little is known was the Yak-16 which was a light twin-engine low-wing cantilever monoplane for 10 passengers.

The wing was of all-metal stressed-skin construction, tapered in chord and thickness and had dihedral outboard of the engines. There were wide-span ailerons and trailing edge split flaps. The fuselage was an oval-section all-metal monocoque structure. All tail surfaces were of metal construction but the rudder was fabric-covered. The single-wheel main undercarriage units retracted into the engine nacelles but the tailwheel was not retractable.

The engines were two 680/750 hp Shvetsov ASh-21 seven-cylinder air-cooled radials developed from half of the ASh-82. There were close-fitting cowlings with cooling gills, and the two-blade controllable-pitch airscrews had large-diameter spinners. Fuel capacity was 1,800 litres (395 Imp. gal) and there was a 70-litre (15 Imp. gal) oil tank in the centre section.

There were five non-reclining seats on each side of the passenger cabin and beside each was a fairly large rectangular window. There were no luggage racks but there was a coat-hanging rail on the rear bulkhead. The entrance door was aft on the port side and opposite it was a lavatory. A door in the rear bulkhead led to an aft cargo hold which also had a hatch in its starboard side.

The crew compartment had dual controls but only the first pilot's panel had flight instruments. A central pedestal contained throttles, propeller controls and trim wheel. Between the pilots' cabin and the passenger cabin was a small compartment which housed the navigator to starboard and provided baggage stowage space to port.

During 1948 a Yak-16, registered SSSR-I985, was exhibited in Czechoslovakia and Poland, and from 9 to 18 June, 1948, it was on show, in company with an Il-12, at Malmi Airport, Helsinki. It is believed that attempts were also made to sell the type in Hungary and Yugoslavia.

The Yak-16 SSSR-I985 on exhibition in Prague in 1948.

Another view of the Yak-16 SSSR-I985.

It has been reported that Yak-16s were used by Aeroflot on local services but there is no evidence for this. One aircraft bore the Aeroflot registration SSSR-L4590 and this may have been used for service trials. It appears that a small number was used by the Soviet Air Force for communications work and as navigational trainers, with the latter having transparent noses and a number of cabin-top astrodomes.

Span 20 m (65 ft 7½ in); length 14·5 m (47 ft 7 in); height 3·6 m (11 ft 10 in). Empty weight 5,000–5,200 kg (11,023–11,464 lb); maximum payload 1,360 kg (2,998 lb); take-off weight 6,400 kg (14,109 lb). Maximum speed 310 km/h (167·28 kt) (192·62 mph); cruising speed at 1,700 m (5,577 ft) 290 km/h (156·48 kt) (180·2 mph); landing speed 85 km/h (45·86 kt) (52·81 mph); take-off run with 25 deg flap 260 m (853 ft); ceiling about 5,000 m (16,404 ft); single-engine ceiling 2,300 m (7,545 ft); optimum range 1,000 km (539 n.miles).

The Yak-16 at Malmi Airport, Helsinki, in June 1948.

Yakovlev Yak-18T

Since 1946 the Yak-18 series of single-engine monoplanes have been the standard primary trainers of the Soviet Air Force. These aircraft have been built in large numbers and supplied to numerous countries, and have, in their more recent versions, achieved notable successes in aerobatic competitions.

The original aircraft was powered by a 160 hp M-11FR five-cylinder air-cooled radial engine, had two seats in tandem and only a semi-retractable tailwheel undercarriage. The Yak-18A was an improved version with 260 hp Ivchenko AI-14R nine-cylinder air-cooled radial engine. The Yak-18U

The unregistered Yak-18T light transport at the 1967 Paris Aero Show. This aircraft was white with red stripes across the rudder. In the background is the An-24TV. (*R. A. Cole.*)

was a version of the original design but with lengthened nose and nosewheel undercarriage. The Yak-18P was a single-seat advanced trainer and aerobatic aircraft with retractable nosewheel undercarriage and 260 hp AI-14R engine. Latest in this line of training and aerobatic aircraft is the Yak-18PM with major modifications including a much further-aft position for the single-seat cockpit, and the 300 hp AI-14RF engine.

In 1966 Sergei Yakovlev, son of the more famous Aleksandr Sergeevich Yakovlev, produced a Yak-18 development design for Aeroflot. This was the Yak-18T cabin monoplane which was first seen in public at the 1967 Paris Aero Show at Le Bourget. Five basic variants of the design visualize its use as a primary trainer; an advanced trainer; a passenger aircraft for pilot, three passengers and baggage; an ambulance capable of taking one stretcher and a medical attendant; and a mail and cargo aircraft with payload of up to 250 kg (551 lb). The aircraft shown at Paris was unregistered, it was not demonstrated and it is not known whether it has completed its flight trials.

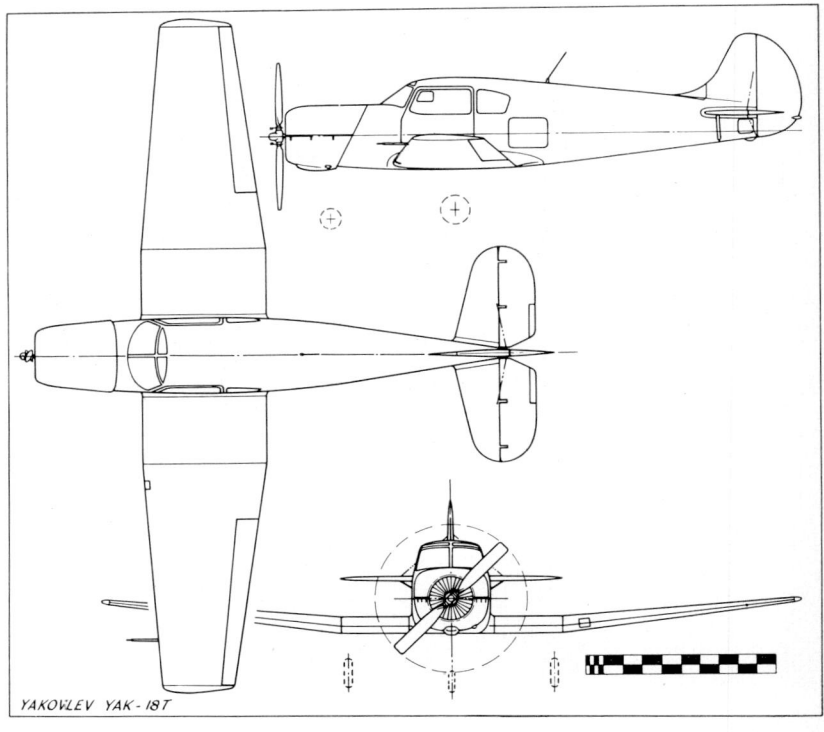

YAKOVLEV YAK-18T

The Yak-18T is a low-wing cantilever monoplane with metal structure, single fin and rudder, and fully-retractable nosewheel undercarriage.

The two-spar wing comprises a centre section of parallel chord and thickness and tapered outer sections with dihedral. The ailerons are inset and without tabs. Centre-section trailing edge flaps are believed to be fitted. There are landing lights in the leading edge of the port wing just outboard of the centre section. The fuselage is of almost rectangular section. The cabin has two individual seats in front and a bench-type seat aft for two passengers. Dual control can be fitted. There is a large forward-opening door on each side, and aft of the cabin is a baggage compartment with an upward-opening door on the port side. The Yak-18T has full radio equipment and ILS.

The fin is smoothly faired into the fuselage, the tailplane is wire braced and the elevators have inset trim tabs. The main undercarriage is inward retracting and the nosewheel retracts aft. All units have single wheels.

The engine is a closely-cowled 300 hp Ivchenko AI-14RF nine-cylinder air-cooled radial with large-diameter multi-blade cooling fan. There is an exhaust collector ring behind the engine and exhaust is ejected through the underside of the cowling. The airscrew is a two-blade variable-pitch V530-D-35.

The Yak-18T standing in the shadow of the An-22 at the 1967 Paris Aero Show. *(M. J. Hooks.)*

Span 11·16 m (36 ft 7¼ in); length 8·354 m (27 ft 5 in); wing area 18·8 sq m (202·36 sq ft). Empty weight 1,200 kg (2,645 lb); maximum payload 250 kg (551 lb); maximum take-off weight 1,620 kg (3,571 lb). Maximum speed 300 km/h (161·88 kt) (186·4 mph); take-off run 200 m (656 ft); landing run 250 m (820 ft); ceiling 5,000 m (16,404 ft); range 1,000 km (539 n.miles).

Two early Yak-24s flying at Tushino in 1955. (*Courtesy William Green.*)

Yakovlev Yak-24

In 1948 A. S. Yakovlev produced his Yak-100 single-engine single-rotor helicopter which was powered by a 420/575 hp AI-26GRFL engine and closely resembled the Sikorsky S-51, but when at the end of the summer of 1951 he was called to a Kremlin meeting to discuss design and production of a twin-engine 24-passenger helicopter it was something of a challenge and a venture into the unknown. Yakovlev discussed the new project with his design bureau staff including Nikolai Skrzhinsky, Pyotr Samsonov, Leon Shekhter and Igor Erlich, and it was decided to accept the challenge but to seek extension of the one year they had been allowed for design, construction and testing. However, they failed to gain an extension of time but the project went ahead.

A tandem-rotor layout was chosen, and Yakovlev has said that they never regretted this decision. At the same time as Yakovlev was asked to produce his twin-engine design, M. L. Mil was commissioned to design and build his single-engine single-rotor Mi-4. The Shvetsov ASh-82V engine was selected to power both types. It is believed that the 21 m (68 ft 10¾ in) diameter rotor was common to both designs except that Yakovlev's design had contra-rotation, with consequent reversal of blades on one rotor.

Yakovlev's design was the Yak-24, on similar lines to the twin-rotor designs of Bristol and Piasecki; and the Soviet aircraft also encountered similar troubles to those experienced with the British and American types. Engine cooling proved to be something of a problem, but this was nothing to the nightmare vibration troubles.

Rotor and power plant systems were built and tested and construction began of four prototype airframes. Two helicopters were built for static and dynamic tests, the third example was for factory tests and the fourth for State trials. The static tests were successfully completed but the dynamic tests failed and the design had to undergo several years of improvements before these were really satisfactory.

The third Yak-24 was completed and after only a few hours running of engines and rotors the vibration problems began to show themselves. A 300-hr endurance test had to be completed before the start of tethered flight tests, and after about 150 hr the vibration troubles appeared to have been more or less cured; but after 178 hr disaster came, when the rear engine mounting collapsed and allowed the rotor to chop the aircraft to pieces, the severed fuel lines completing the destruction by causing the helicopter to catch fire.

Disheartened but undaunted Yakovlev and his team continued work on the fourth aircraft, and test pilots Sergei Brovtsev and Yegor Milyutichev were chosen to make the first free flights, the first taking place on 3 July, 1952. About a hundred short flights, each of a few minutes' duration, were made with the engines throttled to half power. Then a 15 min flight was made and the pilots reported some vibration at certain flight attitudes. Further tests showed not slight vibration but vibration so formidable that it was dangerous.

An unidentified version of the Yak-24. This variant has the metal-covered fuselage and twin vertical fins of the Yak-24U, but appears to closely resemble the later passenger-carrying Yak-24A. This particular Yak-24 was used in 1959 to replace roof trusses of Katherine's Palace, Leningrad. (*Courtesy Aviation Magazine International.*)

This photograph of the Yak-24 was taken while it was lifting roof trusses in Leningrad.

After a thorough investigation of all the possible causes of the vibration it was found that most of it came from the rotor blades; Yakovlev decided to cut 50 cm (19·68 in) off each blade tip whereupon the vibration was cured.

At the start of the winter 1953–54 the Yak-24 was delivered for State trials, but that was not the end of its troubles. After making a few flights in the hands of air force pilots, the aircraft was being run at full power while tethered and with only an engineer in the control cabin. One after the other the guy-ropes snapped and the helicopter promptly took-off. At a height of about 8 m (26 ft) the engineer cut the power and the helicopter canted over and crashed, to become a complete write off.

Finally the Yak-24 design was cleared through its State trials and quantity production was ordered although further work was carried out,

The commercial 30-passenger Yak-24A, bearing Aeroflot's name on the fuselage.
(*Courtesy M. J. Hooks from a photograph at the 1961 Soviet Exhibition, London.*)

particularly to improve the controls. In spite of all the problems involved, production aircraft were flying less than three years from inception of the design and four Yak-24s were seen in public for the first time at Tushino in 1955.

The Yak-24 had a rectangular-section fuselage of steel-tube construction with metal skin to the front and rear sections but with a fabric-covered centre section. The front rotor was mounted above the control cabin and the rear rotor on top of the large tail fin. The front engine air intake was in the front of the shallow pylon on which the rotor was mounted, and the rear engine's air supply came from an intake on each side of the base of the fin leading edge. The rear fuselage and fin trailing edge was twisted to starboard. There were two braced tailplanes with very marked dihedral. The undercarriage comprised four castoring wheels all of the same diameter.

The short-fuselage Yak-24K, with large cabin windows.

The ASh-82V fourteen-cylinder two-row air-cooled radial engines each had a nominal power of 1,430 hp and developed 1,700 hp for take-off. The forward engine was installed behind the control cabin and at an angle. It drove the forward rotor through gearing while the aft engine, installed horizontally, drove the rear rotor via gearing and a shaft inside the fin. The two rotors were interconnected and could be driven from either engine. The tapered rotor blades had tubular steel spars and were metal skinned.

A passageway led from the control cabin to the main cabin and also gave access to the forward engine. The main cabin measured 10 m (32 ft 9½ in) in length and was approximately 2 m (6 ft 6¾ in) in height and width. The Yak-24 could carry up to 4,000 kg (8,818 lb) of cargo or nearly 40 passengers. There was a power-operated downward-opening rear loading

Model of the twin-turbine 39-passenger Yak-24P shown at the Soviet Exhibition in London in 1961. (*Flight International*).

ramp up which vehicles could be driven aboard. An under-fuselage hook made it possible to use the Yak-24 as a crane. Normal crew consisted of two pilots, flight engineer and radio operator.

On 17 December, 1955, a Yak-24 secured two FAI records when Yegor

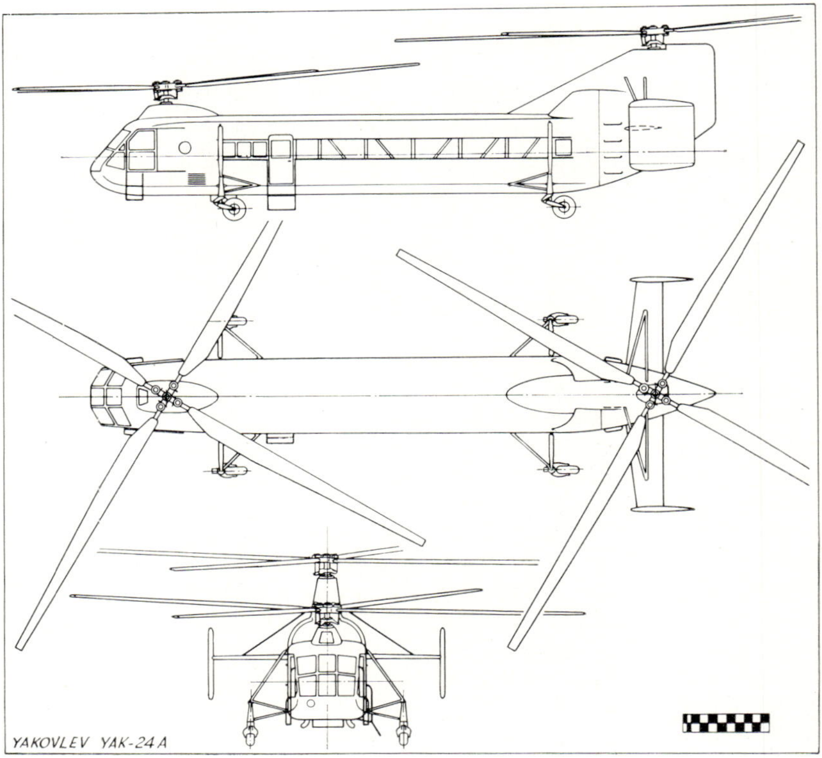

The cut-away model of the Yak-24P shown in London in 1961. (*Flight International.*)

Yak-24P model showing rear of passenger cabin, lavatory, coat-hanging area, baggage hold, rear rotor head and tail unit. (*Flight International.*)

Milyutichev took it to 2,902 m (9,521 ft) with a 4,000 kg (8,818 lb) load, and G. A. Tinyakov climbed to 5,082 m (16,673 ft) with a 2,000 kg (4,409 lb) load. Another pilot, U. A. Garnayev, flew a Yak-24 nonstop from Moscow to Leningrad.

The Yak-24 entered service with the Soviet Air Force but the number used is not known.

In December 1957 the Yak-24U appeared. This was a modified version of the original helicopter but had 21 m (68 ft 10¾ in) diameter rotors, all-metal fuselage, and a horizontal strut-braced tailplane with rectangular end-plate vertical fins. The width of the cabin was also increased by 40 cm (15¾ in) and the empty and loaded weights were increased. There is some evidence that a twin-finned tail layout had earlier been tried on a standard Yak-24 but that its braced tailplane had dihedral.

A 1960 development of the Yak-24U was the Yak-24A designed as a 30-passenger civil transport. Although still powered by the same ASh-82V engines, the Yak-24A had a further increased all-up weight. The seats were in 10 rows with double seats on the starboard side and single units on the port side. Large windows extended the full length of the cabin and the entrance door was forward on the port side. One example of the Yak-24A was certainly flown, and publicity material was produced, but there is no evidence that the type was put into production although it is thought that Aeroflot conducted trials with one of the earlier versions in 1958, and Aeroflot's name did appear above the cabin windows of the Yak-24A.

Also announced in 1960 was the Yak-24K with shortened fuselage and de luxe seating for nine passengers. This version had auto-stabilization, electrically-operated airstairs and four very large cabin windows each side. One example is known to have been built but it is doubtful if this variant ever went into service.

At the Soviet Exhibition in Earl's Court, London, in the summer of 1961, models were displayed of the Yak-24P. This was a project for a 39-passenger development of the Yak-24, with two shaft-turbines said to be of Ivchenko design. The forward turbine was shown as mounted above the forward end of the cabin and the aft turbine was just forward of the vertical fin. Passenger accommodation was arranged in 13 rows of seats, with double units to starboard and single units to port. The entrance door was forward of the passenger cabin on the port side, and aft of the cabin was a lavatory, coat cupboard and a baggage hold. There were seven large windows in each side of the cabin. Nothing has since been heard of this Yak-24P project and it is extremely unlikely that it was ever built.

Yak-24

Rotor diameter 20 m (65 ft 7½ in); overall length with rotors running 33·03 m (108 ft 4½ in); fuselage length 21·34 m (70 ft); height to top of rear rotor head 6·5 m (21 ft 4 in); track 5 m (16 ft 5 in). Empty weight 10,607 kg (23,384 lb); take-off weight 14,270 kg (31,460 lb). Maximum speed 175 km/h (94·43 kt) (108·74 mph); service ceiling 4,200 m (13,779 ft); hover ceiling 2,000 m (6,561 ft); range 265 km (143 n.miles).

Yak-24U

Rotor diameter 21 m (68 ft 10¾ in); overall length with rotors running 34·03 m (111 ft 7¾ in). Other dimensions as Yak-24. Empty weight 11,000 kg (24,251 lb); take-off weight 15,830 kg (34,898 lb). Maximum speed 175 km/h (94·43 kt) (108·74 mph); service ceiling 2,700 m (8,858 ft); hover ceiling 1,500 m (4,921 ft); range 265 km (143 n.miles).

Yak-24A

Rotor diameter 21 m (68 ft 10¾ in); overall length with rotors running 40 m (131 ft 2¾ in). Take-off weight 16,000 kg (35,274 lb). Maximum speed 175 km/h (94·43 kt) (108·74 mph); cruising speed 155 km/h (83·63 kt) (96·31 mph); cruise level 500–1,000 m (1,640–3,280 ft); range with 30 passengers and 300 kg (661 lb) of baggage 200 km (107 n.miles).

Yak-24P

Maximum speed 210 km/h (113·32 kt) (130·49 mph); maximum cruising speed 180 km/h (97·13 kt) (111·85 mph); range 300 km (161 n.miles).

The Yak-40 SSSR-1967 taking part in the Domodedovo display on 9 July, 1967.

Yakovlev Yak-40

About half of Aeroflot's passenger traffic is carried on local services and a high percentage of the places served by these have small aerodromes, many of which do not have paved runways. In 1953, when Aeroflot took steps to re-equip with turbine-powered aircraft, it was only natural that the first phase was devoted to equipment for high-density trunk routes and international operations. Thus from 1956 the turbojet Tu-104s began to appear on main routes and in 1959 the propeller-turbine An-10s and Il-18s went into service. Next came the short-haul An-24s and Tu-124s, but their introduction still left a large network of local services being operated by piston-engined Il-12s, Il-14s, Li-2s and the smaller single-engine An-2s and Yak-12s.

The Yak-40s SSSR-19661 and SSSR-19672.

It was decided that these elderly but useful aircraft should be replaced by a turbojet aircraft offering much greater comfort, double the cruising speed and having the ability to operate from small aerodromes and in poor visual conditions.

Provision of such an aeroplane was by no means easy, as is amply shown by the failure of western countries to produce a suitable aircraft in this category in response to the FAA's design competition in the United States. However, A. S. Yakovlev's design bureau was given the job of meeting Aeroflot's requirements, and the result was the Yak-40 which first flew on 21 October, 1966, as the world's first jet transport designed to operate from short unpaved runways. Work on the project is thought to have started in 1965.

An aircraft using small grass fields has to have particularly good take-off and landing characteristics and so a straight-wing layout with low wing loading was chosen. Safety margins, too, had to be high and for this reason a three-engine design was produced.

A large number of aerodromes in the Soviet Union have only a grass or earth surface and provide a maximum run of only 1,000 m (3,280 ft). In order to serve as many routes as possible the Yak-40 was designed to operate safely from strips measuring only 700 m (2,296 ft) and the take-off distance to 10 m (32 ft) at sea level in ISA is only 400–450 m (1,312–1,476 ft), the actual take-off run being as little as 340–360 m (1,115–1,181 ft).

In layout the Yak-40 is a low-wing cantilever monoplane, with three rear-mounted engines, high T-tail and fully-retractable nosewheel undercarriage. The wing has slight taper on leading and trailing edges and dihedral is $5\frac{1}{2}$ deg. Profiles are S-9 and VK-4 and thickness/chord ratio is

15 per cent at the roots and 10 per cent at the tips. The main spar is a built-up box structure and the wing skin is of varying thickness and chemically etched. There is no centre section, the two wing halves being joined at the centreline, and there are large wing/fuselage fillets. The area between the front spar and the wing nose wall, from aileron to aileron except beneath the fuselage, forms an integral fuel tank. The entire trailing edge is occupied by ailerons and hydraulically-operated slotted-flaps which are in three sections each side. Aileron span is 3·93 m (12 ft 10¾ in) and total aileron area 3·7 sq m (39·82 sq ft). Total flap area is 16·5 sq m (177·6 sq ft), flap setting is 15 deg for take-off and 35 deg for landing. There are no airbrakes or lift-dumpers. Hot-air leading edge deicing is provided.

The fuselage is a circular-section semi-monocoque structure, the skin being attached to the frames and stringers by spot welding and then bonded with a resin glue and finally flush riveted. Workmanship is of a high standard and the whole fuselage is extremely clean aerodynamically. Volume of the pressurized area is 50 cu m (1,765·73 cu ft).

The fin has 52 deg sweepback at 25 per cent chord, the rudder is a single-piece unit with inset tab. The tailplane and elevators are mounted above the fin and are tapered with 15 deg 21 min leading edge sweep, have a span of 7·5 m (23 ft 7¼ in), a thickness/chord ratio of 10 per cent and a total area of 13·03 sq m (140·25 sq ft). The movable tailplane is electrohydraulically-actuated, has a deflection range of 6 deg in each direction and a rate of movement of 0·5 deg/sec.

The undercarriage comprises inward-retracting main units and forward-retracting nosewheel all with single wheels. All units have levered

The Yak-40 SSSR-1967 at Moscow's Sheremetyevo Airport. The aircraft is painted white overall with dark blue fuselage stripe, engine nacelles and fin and rudder. Aeroflot over the cabin windows is in red.

suspension and oleo-pneumatic (nitrogen) shock absorbers. Retraction and extension is hydraulic and there are electrical and mechanical warning systems. When retracted, the main wheels are housed beneath the fuselage, protrude slightly and are not covered by doors. Main-wheel tyre pressure is 4 kg/sq cm (56·89 lb/sq in) and nosewheel tyre pressure 3·5 kg/sq cm (49·78 lb/sq in). The steerable nosewheel has 55 deg deflection from neutral for taxi-ing and 6 deg during take-off and landing.

The engines are three 1,500 kg (3,306 lb) thrust Ivchenko AI-25 turbofans with a specific weight of 0·2 kg per kg of thrust. The AI-25 is a twin-shaft engine with two-spool axial compressors and three-stage turbine. The side engines are carried on short horizontal pylons and have their intakes slightly forward of the wing root trailing edges. The centre engine is mounted in the extreme rear of the fuselage and receives its air via a duct from an intake above the fuselage and forward of the fin. Aft of the central air intake and just forward of the fin is an AI-9 turbostarter. The engines are started from an SV-25 compressed-air unit mounted on each engine and these are spun-up by air bled from the AI-9. The engines can be started from a ground power unit via a connection accessible through the under-fuselage hatch beneath the centre engine. Fuel from the two integral tanks is supplied to the engines by two booster pumps, there is a fuel equalizer to ensure uniformity of flow from each tank and the whole

Yak-40s SSSR-19672, 1967 and 19661. In the distance is a Tu-114.

system of supplying three engines from two tanks was thoroughly rig-tested. There are neither noise suppressors nor thrust reversers.

The first prototype was fitted with more than a ton of test equipment covering more than 1,000 parameters. Other prototypes were used to test the engine installation, deicing system, air-conditioning and other services and special equipment. Design life of the aircraft is 25,000 hr.

The passenger accommodation in the Yak-40 occupies the front half of the fuselage and behind this, on the starboard side, is a large compartment for baggage stowage and, right aft, a lavatory. On the port side is more baggage space and a wardrobe. A door in the rear pressure bulkhead gives access to ventral airstairs. The passenger cabin area is 6·7 m (21 ft 11¾ in) long, 2·26 m (7 ft 5 in) wide and 1·85 m (6 ft 0¾ in) high. Floor area is 10·8 sq m (116·25 sq ft). The standard layout is for 24 passengers, with eight rows of seats at 755 mm (29·7 in) pitch arranged in double units on the starboard side and single units to port. Seating can be increased to 31 by installing double seats on the port side in all but the back row, seat pitch

The Yak-40 SSSR-19661 at the 1967 Paris Aero Show. Looming in the background is the An-22. (*John Stroud.*)

remaining the same. There is a circular window lined up with each seat row and the rear window on each side is incorporated in an overwing emergency exit. All seats can be folded and secured to the cabin walls to provide space for 2,500 kg (5,511 lb) of cargo.

Pressurization provides a cabin altitude of 2,400 m (7,874 ft) at cruise levels of 4,000–6,000 m (13,123–19,685 ft).

The ventral stair door measures 1·64 m by 1 m (5 ft 4½ in by 3 ft 3¼ in) at the actual stairway, there is a 55 cm by 1·1 m (1 ft 9½ in by 3 ft 7¼ in) crew door forward on the port side and the emergency exits measure 48·5 cm by 75 cm (1 ft 7 in by 2 ft 5½ in).

The flight deck is laid out for two-crew operation and the warm air supply for the flight deck is used both for warming the pilots' feet and for heating the windows. Sufficient oxygen is carried for one pilot to remain on oxygen at all times and to provide the second pilot with a supply in case of cabin decompression. Portable oxygen equipment is installed for passenger use and comprises a 1·8 litre (0·39 Imp. gal) bottle, charged to 30 kg/sq cm (426·7 lb/sq in), and masks.

The Yak-40's controls are manually operated, and tailplane, flap and trim tab position indicators and warnings are installed. There are main and emergency hydraulic systems. The main system operates the undercarriage, nosewheel steering, flaps, brakes, windshield wipers, tailplane setting, and ventral stairs; the emergency system allows for operation of the undercarriage, flaps, tailplane, braking and emergency extension and retraction of the airstairs.

A hot-air deicing system is employed for wing, tail and engine intake protection. The centre engine duct is constantly heated with air bled from the air-conditioning system but the other protection can be constant or cyclic. In the latter case the airframe protection is provided in the sequence centre portion of wing, wing tips, fin, and tailplane. Electrical heating is provided for windshields and pitot static heads. There is also a comprehensive fire protection system which covers each engine and the turbostarter.

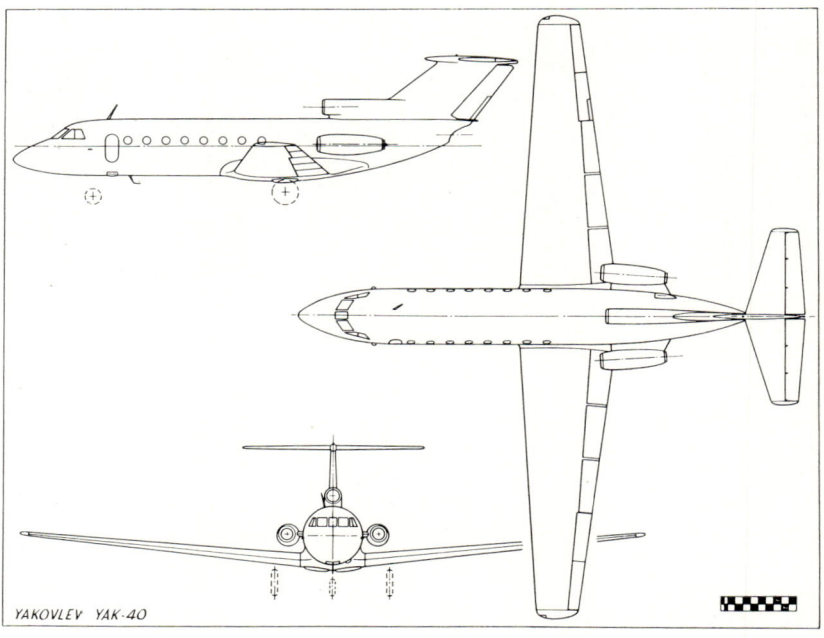

YAKOVLEV YAK-40

'Grosa' weather-warning and map-painting radar is installed, making the Yak-40 the first Soviet local service aircraft to be so equipped. Other navigational and flight equipment includes ARK radio compass, 'Brilliant' radio altimeter, AP-MVL autopilot, two AGB-3K gyro-horizons and SP-50 ILS. Radio and electrical equipment is housed in the nose and aft of the flight deck. The KRP ILS localizer and GRP glide slope aerials are beneath the nose, and the following aerials are in the underside of the fuselage: 'Brilliant-M' radio altimeter, MRP marker receiver, 'Brilliant' altimeter, 'Pero-II' radio and ARK radio compass and two ARK-9 radio compass. The GMK-1G course-setting system is in the wing structure beneath the fuselage. All this equipment is a temporary installation and will later be replaced with miniaturized installations. Automatic approach is possible with minima of 50 m (164 ft) cloud base and 500 m (546 yd) visibility.

The first Yak-40 was registered SSSR-1966 and the second prototype appears to have been SSSR-1967. SSSR-19661 (c/n 019) was exhibited at the 1967 Paris Aero Show, and shortly afterwards photographs were published of SSSR-19671 and SSSR-19672. It seems likely that SSSR-19661 and SSSR-19671 are the original SSSR-1966 and 1967 renumbered. Another Yak-40, seen at Domodedovo in July 1967, was SSSR-19681. SSSR-16540 appeared on a Soviet drawing but this number is known to have been painted on the mock-up and was probably made up from the design year 1965 and the type number Yak-40. Entry into service of the Yak-40 will probably be late in 1968 or some time in 1969. If the type proves successful in Aeroflot service it is almost certain to be built in large numbers and it is known that the Soviet Union is anxious to find export markets for this, so far, unique jet transport.

Span 25 m (82 ft 0¼ in); length 20·19 m (66 ft 3 in); height 6·38 m (20 ft 11¼ in); wing area 70 sq m (753·47 sq ft); track 4·52 m (14 ft 10 in); wheelbase 7·57 m (24 ft 10 in); fuselage ground clearance 1·23 m (4 ft).

Normal fuel 1,800 kg (3,968 lb); maximum fuel 3,000 kg (6,613 lb); normal payload 2,300 kg (5,070 lb); maximum payload 2,800 kg (6,172 lb); normal take-off weight 13,150 kg (28,990 lb); maximum take-off weight 13,700 kg (30,203 lb).

Maximum speed 700 km/h (377·72 kt) (434·96 mph); cruising speed 550–600 km/h (296·78–323·76 kt) (341·75–372·82 mph); take-off and landing run 340–360 m (1,115–1,181 ft); take-off distance to 10 m (32 ft) 400–450 m (1,312–1,476 ft); range at 6,000 m (19,685 ft) and 550 km/h (296·78 kt) (341·75 mph) with 1,800 kg (3,968 lb) fuel including 45 min reserve, 600 km (323 n.miles); range with 3,000 kg (6,613 lb) fuel including 45 min reserve, 1,650 km (890 n.miles).

APPENDIX I

Aeroflot Directorates, Main Bases, Areas of Responsibility, and Equipment— Summer 1967

The Directorate's main base appears after its name.

Figures in parentheses after aircraft types show summer seating capacity on Federal services and, where known, on local services. It has not been possible to distinguish in every case between An-10s and An-10As.

Transliteration of place names into Roman form complies with the agreed system of the Permanent Committee on Geographical Names and the United States Board of Geographic Names.

Directorate	Area	Equipment
Armenian (Yerevan)	Armenian SSR Nakhichevan' ASSR	Federal. An-24V (50), Il-14 (32, and cargo), Il-18 (89) Local. An-2, Il-14
Azerbaydzhan (Baku)	Azerbaydzhan SSR Astrakhan District Dagestan ASSR	Federal. An-24V (50), Il-14 (32), Il-18 (89), Li-2 (21) Local. An-2, Il-14, Li-2, Mi-4. Mi-2 and Mi-8 added later in year
Eastern Siberia (Irkutsk)	Buryat Mongol ASSR Chita District Irkutsk District	Federal. An-10 (85), An-12 (cargo), Il-14 (32), Tu-104B (100), Tu-104D (85) Local. An-2, An-10, An-24, Il-14, Li-2, Mi-4
Estonian (Tallinn)	Estonian SSR	Federal. Il-14 (32), Tu-124 (44), Tu-124V (56) Local. An-2, Il-14, Li-2
Far East (Khabarovsk)	Amur District Khabarovsk Region Primorsk Region Sakhalin District	Federal. Il-14 (32), Il-18 (89), Tu-104A (70), Tu-104D (85) Local. An-2, An-10, Il-14, Il-18, Li-2, Mi-4
Georgian (Tbilisi)	Georgian SSR Abkhaz ASSR Adzhar ASSR	Federal. An-24V (50), Il-14 (28), Tu-104B (100), Tu-104D (85), Tu-124V (56) Local. An-2, Il-14

277

Directorate	Area	Equipment
Kazakh (Alma Ata)	Kazakh SSR	Federal. An-24 (44), Il-14 (32), Il-18 (89), Li-2 (21 and 24) Local. An-2, An-24 (44), Il-14, Il-18, Li-2, Mi-4
Kirgiz (Frunze)	Kirgiz SSR	Federal. Il-14 (32, and cargo), Il-18 (89) Local. An-2, Il-14, Il-18, Mi-4
Komi (Syktyvkar)	Komi ASSR	Federal. An-10A (100), Il-14 (28, 32, 36) Local. An-2, Il-14, Li-2
Krasnoyarsk (Krasnoyarsk)	Krasnoyarsk Region Kemerovo District	Federal. An-24 (48), Il-14, Il-18 (89) Local. An-2, An-24, Il-14, Li-2
Latvian (Riga)	Latvian SSR	Federal. An-24V (50), Il-14 (36), Il-18 (89 and 110) Local. An-2, Il-14, Li-2
Lithuanian (Vil'nyus)	Lithuanian SSR	Federal. Il-14 (32 and 36), Li-2 (cargo), Tu-124 (44), Tu-124V (56) Local. An-2, Il-14, Li-2
Magadan (Magadan)	Magadan District	Federal. An-12 (cargo), Il-14 (28) Local. Il-14
Moldavian (Kishinev)	Moldavian SSR	Federal. An-10A (110), Il-14, Li-2 Local. An-2, Il-14, Li-2
Moscow (Bykovo)	Special tasks in Central Russian District (RSFSR)	Federal. An-24 (44), Il-14 (32, and cargo) Local. An-2, An-24, Il-14, Li-2, L-200D Morava
Moscow Transport	Special tasks	Federal. Il-18 (89 and 110), Il-62, Tu-104 (cargo), Tu-104B (100), Tu-114 (170), Tu-124V (51 and 56)
North (Leningrad)	Karelo-Finnish ASSR Leningrad, Arkhangel'sk, Murmansk, Vologda and Novgorod Districts	Federal. An-24V (50), Il-14 (32, 36 and cargo), Il-18 (89), Li-2 (24, and cargo), Tu-104B (105), Tu-124V (56) Local. An-2, An-24, Il-14, Il-18, Li-2, L-200D Morava, Tu-124
Northern Caucasia (Rostov)	Volgograd, Kamensk and Rostov Districts, Stavropol' and Krasnodar Regions	Federal. An-10A (100), An-24 (48), Il-14 (32), Li-2 (cargo), Tu-124V (56) Local. An-2, An-24, Il-14, Li-2, Mi-4
Polar (Moscow)	Special tasks in Polar Region and Antarctica	Federal. An-10, An-12 (cargo), Il-14, Il-18 (89)
Tadzhik (Dushanbe)	Tadzhik SSR	Federal. Il-14, Il-18 (89) Local. An-2, Il-14, Li-2, Po-2
TUMVL (Moscow)	International operations	— An-12 (cargo), Il-18 (89), Il-62, Tu-104A (70), Tu-104B (100), Tu-114 (120), Tu-124V (56). Tu-134 added later in year

Directorate	Area	Equipment
Turkmen (Ashkhabad)	Turkmen SSR	Federal. An-24V (50), Il-14 (32), Il-18 (89), Li-2 Local. An-2, An-24, Il-14, Li-2, Mi-6, Yak-12
Ukrainian (Kiev)	Ukrainian SSR	Federal. An-10 (85), An-10A (110), An-12 (cargo), An-24 (passenger and cargo), Il-14 (cargo), Li-2, Tu-104B (100), Tu-104D (85) Local. Aero 45, An-2, An-10, An-24 (48), Il-14, Li-2, L-200D Morava, Mi-4, Tu-104B
Ural (Sverdlovsk)	Tyumen, Kurgan, Perm, Sverdlovsk, Chelyabinsk and Kirov Districts Udmurt ASSR	Federal. An-24 (48), Il-14, Il-18 (89) Local. An-2, An-12 (cargo), An-24, Il-14, Il-18, Li-2
Uzbek (Tashkent)	Uzbek SSR	Federal. An-24, Il-14, Il-18 (89) Local. An-2, An-24 (48), Il-14, Il-18, Li-2
Volga (Kuybyshev)	Tatar ASSR, Mari ASSR, Chuvash ASSR, Mordovinian ASSR, Bashkir ASSR, and Gor'kiy, Arzamas, Penza, Saratov, Ul'yanovsk, Kuybyshev and Orenburg Districts	Federal. An-10 (85), An-12 (cargo), An-24 (44), Il-14 (32), Li-2 (cargo), Tu-124 (44) Local. An-2, An-24, Il-14, Li-2
Western Siberia (Novosibirsk)	Omsk and Tomsk Districts, Novosibirsk, and Altai Region	Federal. An-24 (44), Il-14 (cargo), Tu-104B (100), Tu-104D (85) Local. An-2, An-24, Il-14, Li-2
White Russian (Minsk)	White Russian SSR and Kaliningrad District	Federal. An-10 (85), An-24V (50), Il-14 (32), Tu-124V (56) Local. An-2, An-10, An-24, Il-14, Li-2
Yakut (Yakutsk)	Yakut ASSR	Federal. Il-14 (28) Local. An-2, Il-14, Li-2, Mi-4
235 Division		An-24 (36), Il-18 (89), Tu-124* (38)

* Possibly Tu-124K.

APPENDIX II

Aeroflot
Equipment Used on International Services—Summer 1967 and Winter 1967–68

Summer 1967

An-12 (cargo services)

Moscow–Riga–Paris (in association with Air France) and Moscow–Tashkent–Karachi–Colombo–Medan–Djakarta

Il-18

Leningrad–Helsinki, Moscow–Beirut–Cairo, Moscow–Belgrade–Algiers–Bamako–Conakry, Moscow–Budapest–Tunis–Algiers–Rabat, Moscow–Cairo–Khartoum, Moscow–Kiev–Sofia, Moscow–Nicosia–Damascus–Baghdad, Moscow–Omsk–Irkutsk–Ulan Bator, Moscow–Sofia, Moscow–Tashkent–Kabul, Moscow–Tashkent–Karachi–Colombo–Djakarta and Moscow–Teheran–Karachi–Rangoon

Tu-104A

Leningrad–Stockholm–Copenhagen, Moscow–Amsterdam, Moscow–Copenhagen, Moscow–London, Moscow–Omsk–Irkutsk–Peking, Moscow–Omsk–Irkutsk–Pyongyang, Moscow–Paris, Moscow–Prague and Moscow–Rome

Tu-104B

Moscow–Berlin and Moscow–Budapest

Tu-114

Moscow–Delhi, Moscow–Murmansk–Havana, Moscow–Montreal, Moscow–Paris and Moscow–Tokyo (in association with Japan Air Lines)

Tu-124

Kiev–Vienna, Moscow–Belgrade, Moscow–Bucharest, Moscow–Helsinki, Moscow–Kiev–Vienna, Moscow–Stockholm, Moscow–Vienna and Moscow–Warsaw

Winter 1967–68

An-12 (cargo services)
Moscow–Riga–Paris (in association with Air France) and Moscow–Tashkent–Karachi–Colombo–Medan–Djakarta

Il-18
Leningrad–Helsinki, Moscow–Beirut–Cairo, Moscow–Belgrade–Algiers–Bamako–Conakry, Moscow–Bucharest–Sofia, Moscow–Budapest, Moscow–Budapest–Tunis–Algiers–Rabat, Moscow–Cairo–Khartoum, Moscow–Kiev–Bucharest–Sofia, Moscow–Nicosia–Damascus–Baghdad, Moscow–Omsk–Irkutsk–Ulan Bator, Moscow–Sofia, Moscow–Tashkent–Kabul, Moscow–Tashkent–Karachi–Colombo–Djakarta and Moscow–Teheran–Karachi–Rangoon

Il-62
Moscow–Delhi, Moscow–Montreal, Moscow–Rome and Moscow–Paris

Tu-104A
Leningrad–Stockholm–Copenhagen, Moscow–Amsterdam–Brussels, Moscow–Copenhagen, Moscow–London, Moscow–Omsk–Irkutsk–Peking, Moscow–Omsk–Irkutsk–Pyongyang, Moscow–Paris, Moscow–Prague and Moscow–Vienna–Zürich

Tu-104B
Moscow–Berlin

Tu-114
Moscow–Murmansk–Havana, Moscow–Montreal and Moscow–Tokyo (in association with Japan Air Lines)

Tu-134
Moscow–Belgrade, Moscow–Helsinki, Moscow–Kiev–Vienna, Moscow–Stockholm, Moscow–Warsaw and Moscow–Zürich.

APPENDIX III

Aeroflot Equipment Used on Soviet Federal Services— Summer 1966

(Transliteration of place names into Roman form complies with the agreed system of the Permanent Committee on Geographical Names and the United States Board of Geographic Names)

An-10
Eastern Siberia Directorate Irkutsk–Krasnoyarsk, Irkutsk–Mirnyy and Irkutsk–Yakutsk
Ukrainian Directorate Kiev–Adler/Sochi, Leningrad–L'vov (Lwow), L'vov–Odessa–Adler/Sochi and Moscow–Khar'kov
Volga Directorate Kuybyshev–Adler/Sochi–Rostov–Kuybyshev, Kuybyshev–Gor'kiy–Leningrad, Kuybyshev–Kazan'–Leningrad, Kuybyshev–Kazan'–Perm, Kuybyshev–Khar'kov–Kiev, Kuybyshev–Omsk–Novosibirsk, Kuybyshev–Tashkent–Orenburg–Kuybyshev, Kuybyshev–Ufa–Sverdlovsk, Kuybyshev–Volgograd–Rostov–Krasnodar, Moscow–Kuybyshev, Moscow–Orenburg, Moscow–Ufa, Ufa–Adler/Sochi and Ufa–Orenburg–Mineral'nyye Vody–Adler/Sochi
White Russian Directorate Kaliningrad–Minsk–Odessa–Adler/Sochi, Minsk–Gor'kiy–Sverdlovsk, Minsk–Odessa, Minsk–Simferopol, Moscow–Kaliningrad and Moscow–Minsk

An-10A
Komi Directorate Moscow–Syktyvkar, Moscow–Syktyvkar–Noril'sk, Moscow–Syktyvkar–Ukhta, Syktyvkar–Gor'kiy–Krasnodar–Adler/Sochi, Syktyvkar–Gor'kiy–Mineral'nyye Vody, Syktyvkar–Leningrad and Syktyvkar–Sverdlovsk
Moldavian Directorate Kishinev–Donetsk–Kuybyshev–Sverdlovsk, Kishinev–Krasnodar–Mineral'nyye Vody, Kishinev–Leningrad, Kishinev–Minsk–Leningrad, Kishinev–Rostov–Baku, Kishinev–Simferopol, Kishinev–Simferopol–Adler/Sochi and Moscow–Kishinev
Northern Caucasia Directorate Anapa–Kuybyshev–Sverdlovsk–Kuybyshev–Rostov–Anapa, Anapa–Leningrad, Novosibirsk–Omsk–Ufa–Rostov–Adler/Sochi, Moscow–Anapa, Moscow–Krasnodar, Moscow–Rostov, Moscow–Rostov–Groznyy–Moscow, Rostov–Leningrad and Rostov–Simferopol
Ukrainian Directorate Dnepropetrovsk–Adler/Sochi, Dnepropetrovsk–Minsk–Leningrad, Donetsk–Adler/Sochi, Donetsk–Leningrad–Kiev–Donetsk, Khar'kov–Adler/Sochi, Khar'kov–Leningrad, Kherson–Leningrad, Kiev–Donetsk–Adler/Sochi, Kiev–Khar'kov–Kazan'–Sverdlovsk, Lugansk–Adler/Sochi, Moscow–Dnepropetrovsk, Moscow–Donetsk, Moscow–Khar'kov, Moscow–Kiev–Kherson, Moscow–Krivoy Rog, Moscow–L'vov, Moscow–Lugansk, Moscow–Nikolayev, Nikolayev–Leningrad, Odessa–Khar'kov–Ufa–Sverdlovsk and Zaporozh'ye–Adler/Sochi

An-12 (cargo services)
Eastern Siberia Directorate Irkutsk–Mirnyy
Polar Directorate Leningrad–Gor'kiy–Chelyabinsk–Novosibirsk–Chelyabinsk–Moscow, Moscow–Minsk–Gor'kiy–Sverdlovsk–Novosibirsk–Chelyabinsk–Gor'kiy–Leningrad and Moscow–Sverdlovsk–Novosibirsk–Krasnoyarsk–Irkutsk–Blagoveshchensk–Khabarovsk
Ukrainian Directorate Khar'kov–Perm–Novosibirsk
Volga Directorate Gor'kiy–Kuybyshev–Omsk–Krasnoyarsk–Omsk–Gor'kiy, Gor'kiy–Omsk–Novosibirsk, Kazan'–Omsk–Novosibirsk, Kuybyshev–Khar'kov, Kuybyshev–Omsk–Krasnoyarsk and Kuybyshev–Ufa–Novosibirsk–Omsk–Ufa–Kuybyshev

An-24
Kazakh Directorate Alma Ata–Karaganda–Tselinograd–Kokchetav–Petropavlovsk–Kurgan–Sverdlovsk, Moscow–Penza–Ural'sk–Aktyubinsk–Dzhezkazgan–Balkhash–Alma Ata and Tselinograd–Karaganda
Krasnoyarsk Directorate Krasnoyarsk–Novosibirsk
Latvian Directorate Riga–Liepaya–Kaliningrad, Riga–Minsk–L'vov–Kishinev, Riga–Tallinn and Riga –Vil'nyus (Vilna)–L'vov–Kishinev
Moscow Directorate Ivanovo–Voronezh–Rostov–Adler/Sochi, Kursk–Donetsk–Gudauta, Lipetsk–Rostov–Adler/Sochi, Moscow–Cheboksary, Moscow–Ivanovo, Moscow–Kazan'–Izhevsk, Moscow–Kazan'–Tyumen, Moscow–Kiev–Ivano-Frankovsk, Moscow–Kursk, Moscow–Lipetsk, Moscow–Penza, Moscow–Penza–Ural'sk–Gur'yev, Moscow–Poltava–Kirovograd, Moscow–Saratov, Moscow–Tambov, Moscow–Tula, Moscow–Voronezh, Moscow–Voronezh–Rostov–Nal'chik, Moscow–Voronezh–Rostov–Ordzhonikidze, Moscow–Voronezh–Volgograd–Elista-Groznyy, Moscow–Voronezh–Volgograd–Makhachkala, Tambov–Rostov–Gudauta, Tula–Donetsk–Gudauta, Voronezh–Krasnodar–Adler/Sochi and Voronezh–Simferopol
Northern Caucasia Directorate Moscow–Voronezh–Rostov–Gelendzhik
Turkmen Directorate Ashkhabad–Mary–Samarkand–Tashkent and Ashkhabad–Nebit-Dag–Baku
Ural Directorate Moscow–Kazan'–Sverdlovsk–Kurgan and Tyumen–Sverdlovsk
Volga Directorate Penza–Volgograd–Mineral'nyye Vody–Gudauta, Saratov–Khar'kov–Simferopol, Saratov–Kislovodsk, Saratov–Krasnodar–Gudauta–Volgograd–Saratov, Saratov–Kuybyshev–Ufa–Perm, Saratov–Rostov–Adler/Sochi, Saratov–Volgograd–Krasnodar and Saratov–Voronezh–Khar'kov
Western Siberia Directorate Novosibirsk–Kemerovo–Abakan and Novosibirsk–Omsk–Tyumen
235 Division Moscow–Khar'kov–Gelendzhik

An-24 (cargo services)
Ukrainian Directorate Kiev–Donetsk, Kiev–Odessa and Moscow–Kiev

An-24 (known Regional services)
Kazakh Directorate Karaganda–Tselinograd
Uzbek Directorate Tashkent–Bukhara and Tashkent–Samarkand

Il-14
Armenian Directorate Yerevan (Erivan)–Gudauta–Krasnodar–Dnepropetrovsk, Yerevan–Sukhumi–Rostov–Khar'kov and Yerevan–Tbilisi

Il-14 *cont.*

Azerbaydzhan Directorate Astrakhan–Krasnodar–Simferopol, Astrakhan–Mineral'nyye Vody–Adler/Sochi, Astrakhan–Stavropol'–Krasnodar–Anapa, Astrakhan–Volgograd–Saratov–Penza, Baku–Makhachkala–Astrakhan–Aktyubinsk, Baku–Makhachkala–Astrakhan–Saratov, Baku–Urgench–Tashauz, Mineral'nyye Vody–Astrakhan–Ural'sk–Orenburg–Magnitogorsk, Rostov–Elista–Astrakhan and Rostov–Astrakhan
Eastern Siberia Directorate Irkutsk–Bratsk–Krasnoyarsk, Irkutsk–Kirensk–Vitim–Mirnyy–Nyurba and Irkutsk–Kirensk–Vitim–Mukhtuya
Estonian Directorate Tallinn–Kohtla-Järve–Leningrad, Tallinn–Narva–Leningrad, Tallinn–Riga–Liepaya–Kaliningrad and Tallinn–Riga–Vil'nyus–Minsk
Far East Directorate Khabarovsk–Blagoveshchensk–Magdagachi–Takhtamygda–Chita–Irkutsk
Georgian Directorate Batumi–Sukhumi–Rostov–Khar'kov and Tbilisi–Kutaisi–Krasnodar
Kazakh Directorate Aktyubinsk–Gur'yev–Kislovodsk, Aktyubinsk–Orsk–Magnitogorsk–Chelyabinsk–Sverdlovsk, Chimkent–Tashkent, Dzhambul–Chimkent–Tashkent, Kustanay–Aktyubinsk–Gur'yev–Astrakhan–Kislovodsk, Moscow–Penza–Ural'sk–Aktyubinsk and Moscow–Penza–Ural'sk–Aktyubinsk–Kustanay
Kirgiz Directorate Frunze–Karaganda
Komi Directorate Syktyvkar–Arkhangel'sk (Archangel), Syktyvkar–Izhevsk–Sverdlovsk, Syktyvkar–Kirov–Perm, Syktyvkar–Pechora–Inta, Syktyvkar–Ukhta, Syktyvkar–Vorkuta and Vorkuta–Noril'sk
Latvian Directorate Moscow–Riga
Lithuanian Directorate Druskininkay–Kaunas–Riga–Leningrad, Klaypeda (Memel)–Kaunas–Vil'nyus–Minsk, Klaypeda–Riga–Leningrad, Moscow–Vil'nyus–Kaunas–Klaypeda and Vil'nyus–Kiev–Dnepropetrovsk–Krasnodar
Magadan Directorate Khabarovsk–Nikolayevsk–Okhotsk–Magadan
Moscow Directorate Lipetsk–Khar'kov–Simferopol, Moscow–Bryansk–Kiev–Vinnitsa, Moscow–Ioshkar-Ola, Moscow–Kazan'–Magnitogorsk, Moscow–Kiev–Chernovtsy (Cernăuți), Moscow–Kursk–Sumy, Moscow–Kuybyshev–Magnitogorsk–Kazan'–Moscow, Moscow–Voronezh, Voronezh–Dnepropetrovsk–Odessa, Voronezh–Khar'kov–Zaporozh'ye, Voronezh–Kiev–Minsk and Voronezh–Rostov–Kislovodsk
North Directorate Moscow–Cherepovets, Moscow–Cherepovets–Kotlas, Moscow–Gor'kiy–Kirov, Moscow–Novgorod and Moscow–Velikiye Luki–Pskov
Northern Caucasia Directorate Rostov–Elista–Astrakhan
Turkmen Directorate Ashkhabad–Mary–Chardzhou–Tashkent
Volga Directorate Gur'yev–Ural'sk–Kuybyshev, Gor'kiy–Kazan'–Izhevsk–Perm, Kuybyshev–Penza–Voronezh, Moscow–Balakovo, Moscow–Gor'kiy, Moscow–Saransk and Moscow–Saratov
White Russian Directorate Brest–Kiev, Kaliningrad–Minsk, Minsk–Kiev, Moscow–Mogilev and Moscow–Vitebsk
Yakut Directorate Aldan–Chul'man–Takhtamygda–Chita–Irkutsk, Aldan–Chul'man–Takhtamygda–Magdagachi–Blagoveshchensk–Khabarovsk, Irkutsk–Kirensk–Vitim–Olekminsk–Yakutsk and Novosibirsk–Krasnoyarsk–Bratsk–Kirensk–Olekminsk–Yakutsk

Il-14 (cargo services)

Armenian Directorate Moscow–Voronezh–Rostov–Sukhumi–Yerevan
Kirgiz Directorate Moscow–Penza–Ural'sk–Aktyubinsk–Dzhezkazgan–Balkhash–Frunze

Moscow Directorate Moscow–Cheboksary–Izhevsk–Sverdlovsk–Kazan'–Moscow and Moscow–Kuybyshev–Sverdlovsk–Kazan'–Moscow
North Directorate Chelyabinsk–Sverdlovsk–Perm–Kirov–Cherepovets–Leningrad and Moscow–Leningrad
Ukrainian Directorate Kiev–L'vov and Moscow–Khar'kov–Dnepropetrovsk–Zaporozh'ye–Simferopol
Western Siberia Directorate Novosibirsk–Omsk–Sverdlovsk

Il-14 (known Regional services)
Kirgiz Directorate Frunze–Alma Ata
Moldavian Directorate Kishinev–Kiev and Kishinev–Odessa
Western Siberia Directorate Novosibirsk–Barnaul, Novosibirsk–Biysk and Novosibirsk–Tomsk
White Russian Directorate Minsk–Brest and Minsk–Grodno

Il-18
Armenian Directorate Moscow–Yerevan, Yerevan–Adler/Sochi, Yerevan–Adler/Sochi–Rostov–Gor'kiy, Yerevan–Ashkhabad–Tashkent, Yerevan–Krasnodar–Kiev–L'vov, Yerevan–Mineral'nyye Vody, Yerevan–Mineral'nyye Vody–Leningrad, Yerevan–Simferopol, Yerevan–Simferopol–Odessa and Yerevan–Volgograd–Kazan'–Sverdlovsk
Azerbaydzhan Directorate Astrakhan–Krasnodar–Simferopol, Astrakhan–Mineral'nyye Vody–Adler/Sochi, Baku–Ashkhabad–Tashkent, Baku–Astrakhan–Gor'kiy, Baku–Astrakhan–Kazan', Baku–Astrakhan–Kuybyshev–Sverdlovsk, Baku–Astrakhan–Leningrad, Baku–Khar'kov–Minsk, Baku–Krasnodar–Odessa, Baku–Krasnodar–Simferopol, Baku–Krasnodar–Simferopol–Odessa, Baku–Krasnovodsk, Baku–Mineral'nyye Vody–Rostov–Kiev, Baku–Simferopol–Odessa–L'vov, Baku–Volgograd, Baku–Yerevan, Moscow–Astrakhan–Baku and Moscow–Baku
Far East Directorate Khabarovsk–Irkutsk–Novosibirsk–Adler/Sochi, Khabarovsk–Magadan, Moscow–Krasnoyarsk–Khabarovsk–Yuzhno-Sakhalinsk' and Moscow–Krasnoyarsk–Yuzhno-Sakhalinsk'
Kazakh Directorate Alma Ata–Baku–Adler/Sochi, Alma Ata–Semipalatinsk–Novosibirsk, Alma Ata–Tselinograd–Kuybyshev–Kiev, Alma Ata–Tselinograd–Mineral'nyye Vody, Alma Ata–Tselinograd–Mineral'nyye Vody–Adler/Sochi, Alma Ata–Tselinograd–Omsk, Alma Ata–Tselinograd–Rostov–Simferopol, Alma Ata–Tselinograd–Sverdlovsk, Moscow–Alma Ata, Moscow–Omsk–Semipalatinsk, Moscow–Semipalatinsk–Omsk–Moscow, Moscow–Tselinograd, Moscow–Tselinograd–Alma Ata and Semipalatinsk–Tselinograd–Mineral'nyye Vody–Adler/Sochi
Kirgiz Directorate Frunze–Baku–Mineral'nyye Vody, Frunze–Baku–Mineral'nyye Vody–Adler/Sochi, Frunze–Novosibirsk, Frunze–Omsk, Frunze–Semipalatinsk–Kemerovo–Krasnoyarsk, Frunze–Tashkent, Frunze–Tselinograd–Novosibirsk, Moscow–Osh, Moscow–Frunze and Moscow–Frunze–Kuybyshev–Moscow
Krasnoyarsk Directorate Moscow–Chelyabinsk–Krasnoyarsk, Moscow–Kazan'–Krasnoyarsk, Moscow–Kazan'–Krasnoyarsk–Vladivostok, Moscow–Krasnoyarsk, Moscow–Krasnoyarsk–Blagoveshchensk, Moscow–Novosibirsk–Krasnoyarsk–Irkutsk, Moscow–Omsk–Krasnoyarsk–Irkutsk, Moscow–Sverdlovsk–Krasnoyarsk, Moscow–Sverdlovsk–Krasnoyarsk–Irkutsk, Tashkent–Alma Ata–Kemerovo–Krasnoyarsk, Moscow–Alma Ata–Novosibirsk–Krasnoyarsk and Moscow–Alma Ata–Novosibirsk–Krasnoyarsk–Yakutsk–Magadan

Il-18 *cont.*
Latvian Directorate Moscow–Riga, Riga–Donetsk–Tbilisi–Baku, Riga–Gor'kiy–Sverdlovsk–Omsk–Novosibirsk, Riga–Kazan'–Sverdlovsk–Omsk–Novosibirsk, Riga–Khar'kov–Krasnodar–Yerevan, Riga–Khar'kov–Volgograd–Tashkent, Riga–Kiev–Odessa–Adler/Sochi, Riga–Kiev–Odessa–Simferopol–Adler/Sochi, Riga–Kiev–Rostov–Mineral'nyye Vody, Riga–Kiev–Simferopol, Riga–Leningrad and Riga–Leningrad–Arkhangel'sk

Moscow Transport Directorate Moscow–Adler/Sochi, Moscow–Donetsk–Sukhumi, Moscow–Gor'kiy, Moscow–Gudauta, Moscow–Kazan', Moscow–Kemerovo–Bratsk, Moscow–Krasnodar, Moscow–Krasnoyarsk–Magadan, Moscow–Noril'sk, Moscow–Novosibirsk, Moscow–Novosibirsk–Bratsk, Moscow–Novosibirsk–Krasnoyarsk–Irkutsk–Magadan, Moscow–Rostov, Moscow–Volgograd and Moscow–Zaporozh'ye

North Directorate Arkhangel'sk–Kiev–Odessa, Arkhangel'sk–Krasnodar–Adler/Sochi, Arkhangel'sk–Rostov–Mineral'nyye Vody, Arkhangel'sk–Simferopol, Khabarovsk–Irkutsk–Novosibirsk–Kuybyshev, Kuybyshev–Gor'kiy–Leningrad, Leningrad–Adler/Sochi, Leningrad–Arkhangel'sk–Noril'sk, Leningrad–Chelyabinsk–Frunze, Leningrad–Chelyabinsk–Tselinograd–Alma Ata, Leningrad–Gor'kiy–Krasnoyarsk–Khabarovsk–Vladivostok, Leningrad–Gor'kiy–Ufa, Leningrad–Gudauta, Leningrad–Kaliningrad, Leningrad–Kazan'–Tselinograd–Alma Ata, Leningrad–Krasnodar, Leningrad–Lugansk–Mineral'nyye Vody, Leningrad–Omsk–Krasnoyarsk–Chita–Vladivostok, Leningrad–Sverdlovsk–Tashkent, Leningrad–Ufa–Krasnoyarsk–Blagoveshchensk–Vladivostok, Leningrad–Volgograd, Leningrad–Zaporozh'ye–Krasnodar and Moscow–Arkhangel'sk

Polar Directorate Moscow–Khatanga–Tiksi and Moscow–Tiksi–Anadyr'

Tadzhik Directorate Dushanbe–Ashkhabad–Baku–Mineral'nyye Vody, Dushanbe–Ashkhabad–Mineral'nyye Vody, Dushanbe–Baku–Mineral'nyye Vody, Dushanbe–Leninabad–Alma Ata–Novosibirsk, Dushanbe–Leninabad–Chelyabinsk, Dushanbe–Leninabad–Frunze, Dushanbe–Leninabad–Kuybyshev, Dushanbe–Leninabad–Sverdlovsk, Dushanbe–Mineral'nyye Vody–Simferopol–Kiev, Dushanbe–Tashkent, Dushanbe–Tbilisi–Adler/Sochi, Moscow–Dushanbe, Moscow–Kuybyshev–Leninabad–Dushanbe and Moscow–Leninabad–Dushanbe

TUMVL (International) Kiev–Simferopol–Adler/Sochi (connecting with Interflug's Berlin–Kiev services) and Moscow–Adler/Sochi

Turkmen Directorate Ashkhabad–Baku–Tbilisi–Adler/Sochi, Ashkhabad–Krasnodar–Simferopol, Ashkhabad–Krasnovodsk–Astrakhan–Volgograd–Kuybyshev, Ashkhabad–Krasnovodsk–Khar'kov–Leningrad, Ashkhabad–Krasnovodsk–Krasnodar–Simferopol, Ashkhabad–Krasnovodsk–Mineral'nyye Vody, Ashkhabad–Krasnovodsk–Mineral'nyye Vody–Krasnodar–Simferopol, Ashkhabad–Krasnovodsk–Rostov–Kiev, Ashkhabad–Mineral'nyye Vody, Ashkhabad–Tashkent, Ashkhabad–Volgograd–Kuybyshev, Moscow–Ashkhabad and Moscow–Krasnovodsk–Ashkhabad

Ural Directorate Chelyabinsk–Gor'kiy–Leningrad, Chelyabinsk–Kazan'–Leningrad, Chelyabinsk–Krasnodar–Simferopol, Chelyabinsk–Mineral'nyye Vody, Chelyabinsk–Mineral'nyye Vody–Adler/Sochi, Chelyabinsk–Rostov–Simferopol, Kemerovo–Sverdlovsk–Mineral'nyye Vody–Adler/Sochi, Moscow–Chelyabinsk, Moscow–Perm, Moscow–Perm–Sverdlovsk, Moscow–Sverdlovsk–Kemerovo, Perm–Mineral'nyye Vody–Adler/Sochi, Perm–Sverdlovsk–Tashkent, Sverdlovsk–Adler/Sochi, Sverdlovsk–Chelyabinsk–Kiev–L'vov, Sverdlovsk–Gor'kiy–Leningrad, Sverdlovsk–Krasnodar, Sverdlovsk–Krasnodar–Simferopol, Sverdlovsk–Mineral'nyye Vody, Sverdlovsk–Mineral'nyye Vody–Gudauta, Sverdlovsk–Perm–Leningrad, Sverdlovsk–Rostov–Adler/Sochi, Sverdlovsk–Simferopol, Sverdlovsk–Volgograd–Gudauta and Tashkent–Sverdlovsk

Uzbek Directorate Moscow–Nukus–Tashkent, Moscow–Tashkent, Tashkent–Adler/ Sochi, Tashkent–Alma Ata, Tashkent–Chelyabinsk, Tashkent–Krasnodar–Kiev, Tashkent–Mineral'nyye Vody, Tashkent–Novosibirsk, Tashkent–Nukus–Baku–Mineral'nyye Vody, Tashkent–Tbilisi, Tashkent–Tselinograd–Omsk, Tashkent–Simferopol, Tashkent–Simferopol–Odessa and Tashkent–Ufa–Kazan'
235 Division Moscow–Adler/Sochi

Li-2

Azerbaydzhan Directorate Astrakhan–Gur'yev
Kazakh Directorate Astrakhan–Gur'yev, Dzhambul–Tashkent, Gur'yev–Astrakhan–Kislovodsk, Gur'yev–Ural'sk–Kuybyshev, Karaganda–Ekibastuz Ugol'–Pavlodar–Omsk, Kustany–Chelyabinsk–Sverdlovsk and Semipalatinsk–Ust'-Kamenogorsk
Moscow Directorate Kursk–Kiev–Sumy–Kursk, Moscow–Voronezh, Tambov–Lipetsk–Voronezh–Khar'kov and Tambov–Voronezh–Rostov–Kislovodsk
North Directorate Moscow–Vologda–Vel'sk–Berezniki–Arkhangel'sk

Li-2 (cargo services)

Moscow Directorate Ivanovo–Gor'kiy–Kazan'–Chelyabinsk–Kazan'–Gor'kiy–Moscow–Ivanovo, Kursk–Khar'kov–Dnepropetrovsk–Kherson, Kursk–Khar'kov–Donetsk–Rostov, Kursk–Voronezh–Kuybyshev, Moscow–Cheboksary–Ioshkar-Ola, Moscow–Gor'kiy–Kazan'–Izhevsk, Moscow–Gor'kiy–Kazan'–Izhevsk–Perm, Moscow–Kazan'–Ufa–Chelyabinsk–Ufa–Gor'kiy–Moscow, Moscow–Penza–Saratov, Moscow–Voronezh–Volgograd and Voronezh–Khar'kov–Kiev
North Directorate Moscow–Gor'kiy–Kirov
Northern Caucasia Directorate Moscow–Voronezh–Rostov–Krasnodar
Ural Directorate Chelyabinsk–Kurgan–Omsk–Novosibirsk–Tomsk, Moscow–Penza–Kuybyshev–Ufa–Chelyabinsk and Sverdlovsk–Perm–Kazan'
Volga Directorate Moscow–Penza–Saratov
White Russian and Lithuanian Directorates (Joint operation) Moscow–Minsk–Vil'nyus

Li-2 (known Regional services)

Ukrainian Directorate Khar'kov–Dnepropetrovsk and Khar'kov–Donetsk

Tu-104A

Far East Directorate Moscow–Omsk–Irkutsk–Khabarovsk–Petropavlovsk-Kamchatskiy

Tu-104B

Eastern Siberia Directorate Irkutsk–Omsk–Sverdlovsk–Leningrad, Moscow–Chelyabinsk–Novosibirsk–Sverdlovsk–Moscow, Moscow–Omsk–Irkutsk, Moscow–Omsk–Irkutsk–Chita, Moscow–Omsk–Irkutsk–Chita–Irkutsk–Omsk–Sverdlovsk–Moscow and Moscow–Sverdlovsk–Omsk–Irkutsk
Georgian Directorate Moscow–Kutaisi, Moscow–Sukhumi, Moscow–Tbilisi, Tbilisi–Kiev–Leningrad, Tbilisi–Mineral'nyye Vody, Tbilisi–Sukhumi–Simferopol and Tbilisi–Sukhumi–Simferopol–Odessa
Moscow Transport Directorate Moscow–Mineral'nyye Vody, Moscow–Omsk, Moscow–Simferopol, Moscow–Sukhumi, Moscow–Sverdlovsk and Moscow–Ul'yanovsk
North Directorate Leningrad–Kiev, Leningrad–Mineral'nyye Vody, Leningrad–Odessa, Leningrad–Simferopol, Leningrad–Sukhumi, Leningrad–Sverdlovsk–Omsk–Irkutsk–Khabarovsk and Moscow–Leningrad

Tu-104B cont.

Ukrainian Directorate Kiev–Sukhumi, Leningrad–Sukhumi, Moscow–Kiev, Moscow–Odessa, Moscow–Simferopol and Tbilisi–Kiev

Western Siberia Directorate Leningrad–Sverdlovsk–Novosibirsk–Chita–Khabarovsk–Vladivostok, Moscow–Novosibirsk–Sverdlovsk–Moscow, Moscow–Omsk–Novosibirsk–Chelyabinsk–Moscow, Moscow–Sverdlovsk–Novosibirsk–Omsk–Moscow, Novosibirsk–Chita–Khabarovsk–Vladivostok and Novosibirsk–Irkutsk–Khabarovsk–Vladivostok

Tu-104D

Eastern Siberia Directorate Irkutsk–Omsk–Mineral'nyye Vody–Simferopol, Khabarovsk–Irkutsk, Khabarovsk–Irkutsk–Omsk–Sverdlovsk, Moscow–Chelyabinsk–Omsk–Irkutsk, Moscow–Chelyabinsk–Omsk–Irkutsk–Khabarovsk and Moscow–Omsk–Irkutsk

Far East Directorate Khabarovsk–Irkutsk–Semipalatinsk–Alma Ata, Moscow–Omsk–Irkutsk–Khabarovsk–Vladivostok and Petropavlovsk-Kamchatskiy–Khabarovsk–Irkutsk–Omsk–Chelyabinsk–Mineral'nyye Vody–Simferopol

Georgian Directorate Sukhumi–Sverdlovsk and Tbilisi–Sverdlovsk

Ukrainian Directorate Leningrad–Odessa, Leningrad–Simferopol, Moscow–Odessa, Moscow–Simferopol, Odessa–Simferopol–Sverdlovsk–Omsk and Odessa–Sukhumi

Western Siberia Directorate Novosibirsk–Omsk–Chelyabinsk–Leningrad, Novosibirsk–Omsk–Chelyabinsk–Mineral'nyye Vody and Vladivostok–Khabarovsk–Irkutsk–Novosibirsk–Chelyabinsk–Kiev–Odessa

Tu-104* (cargo service)

Moscow Transport Directorate Moscow–Omsk–Irkutsk–Khabarovsk

Tu-114

Moscow Transport Directorate Moscow–Khabarovsk

Tu-124

Estonian Directorate Moscow–Tallinn, Tallinn–Kiev–Adler/Sochi, Tallinn–Kiev–Simferopol and Tallinn–Leningrad

Georgian Directorate Tbilisi–Adler/Sochi, Tbilisi–Kutaisi–Donetsk and Tbilisi–Kutaisi–Rostov

Lithuanian Directorate Moscow–Vil'nyus, Vil'nyus–Donetsk–Adler/Sochi, Vil'nyus–Kiev–Donetsk–Adler/Sochi, Vil'nyus–Kiev–Gudauta, Vil'nyus–Kiev–Odessa, Vil'nyus–Kiev–Rostov–Mineral'nyye Vody, Vil'nyus–Kiev–Simferopol, Vil'nyus–Leningrad and Vil'nyus–Simferopol

Moscow Transport Directorate Moscow–Adler/Sochi, Moscow–Murmansk and Moscow–Stavropol'

North Directorate Moscow–Petrozavodsk and Murmansk–Leningrad

Northern Caucasia Directorate Mineral'nyye Vody–Gor'kiy, Mineral'nyye Vody–Krasnodar–Odessa, Mineral'nyye Vody–Rostov–L'vov, Mineral'nyye Vody–Simferopol, Mineral'nyye Vody–Simferopol–Kiev–Kaliningrad, Mineral'nyye Vody–Volgograd–Ufa, Minsk–Zaporozh'ye–Mineral'nyye Vody, Moscow–Khar'kov–Mineral'nyye Vody, Moscow–Volgograd, Volgograd–Adler/Sochi, Volgograd–Donetsk–L'vov, Volgograd–Gor'kiy, Volgograd–Khar'kov–Minsk, Volgograd–Kiev, Volgograd–Rostov–Odessa and Volgograd–Rostov–Simferopol

* Actual version not known.

TUMVL (International) Moscow–Adler/Sochi
Volga Directorate Gor'kiy–Khar'kov–Simferopol, Kazan'–Adler/Sochi, Kazan'–Khar'kov–Simferopol, Kazan'–Mineral'nyye Vody, Kazan'–Rostov–Krasnodar, Kazan'–Volgograd–Gudauta, Kuybyshev–Adler/Sochi, Kuybyshev–Mineral'nyye Vody, Kuybyshev–Simferopol, Kuybyshev–Tbilisi, Moscow–Kazan', Ufa–Khar'kov–Simferopol, Ufa–Krasnodar–Adler/Sochi and Ul'yanovsk–Mineral'nyye Vody–Adler/Sochi
White Russian Directorate Minsk–Donetsk–Krasnodar, Minsk–Donetsk–Mineral'nyye Vody, Minsk–Leningrad, Minsk–Odessa, Minsk–Simferopol–Adler/Sochi and Moscow–Minsk
235 Division Moscow–Gor'kiy and Moscow–Stavropol'

APPENDIX IV

Routes Operated by Soviet-designed Aircraft in the Service of Non-Soviet Airlines—Summer 1967

(Misrair operations are those prior to the Israeli attack on 5 June)

An-2

Air Mali Bamako–Kenieba, Bamako–Kenieba–Kayes, Bamako–Segou–Mopti–Niafounké–Goundam–Tombouctou–Rharous and Kayes–Yélimané–Nioro
CAAC (China) Lanchow–Sining and Urumchi–Karamai

An-24

Air Guinée Conakry–Boké–Labé–Kankan, Conakry–Dakar, Conakry–Freetown–Monrovia, Conakry–Kankan–Bamako, Conakry–Kankan–Kissidougou–Macenta, Conakry–Kankan–Macenta–N'zérékoré, Conakry–Kankan–N'zérékoré and Conakry–Kissidougou–Kankan–Siguiri
Air Mongol - Mongolian Airlines Ulan Bator–Irkutsk
Cubana Havana–Camaguey and Havana–Holguín
Interflug Berlin–Barth, Berlin–Dresden, Berlin–Erfurt, Berlin–Heringsdorf, Berlin–Leipzig, Berlin–Warsaw, Dresden–Barth, Dresden–Budapest, Dresden–Erfurt, Dresden–Heringsdorf, Leipzig–Barth, Leipzig–Budapest and Leipzig–Heringsdorf
Misrair* Cairo–Abo Redis, Cairo–Alexandria, Cairo–Alexandria–Athens, Cairo–Alexandria–Mersa Matruh, Cairo–El Arish, Cairo–El Arish–Beirut, Cairo–Hurghada, Cairo–Luxor–Aswan, Cairo–New Valley, Cairo–Nicosia, Cairo–Port Said–Athens and Cairo–Port Said–El Arish
Polskie Linie Lotnicze (LOT) Kraków–Budapest, Kraków–Gdańsk (Danzig), Kraków–Gdańsk–Copenhagen, Kraków–Poznań, Warsaw–Belgrade, Warsaw–Budapest, Warsaw–Budapest–Bucharest, Warsaw–Copenhagen, Warsaw–Gdańsk, Warsaw–Katowice, Warsaw–Kiev, Warsaw–Kraków, Warsaw–Poznań, Warsaw–Prague, Warsaw–Stockholm–Helsinki, Warsaw–Vienna, Warsaw–Vil'nyus–Leningrad, Warsaw–Wrocław, Warsaw–Zürich and Wrocław–Gdańsk
Tabso† Sofia–Athens, Sofia–Burgas, Sofia–Rousse, Sofia–Varna and Sofia–Vienna–Zürich

* Now merged with United Arab Airlines
† Tabso was renamed Bulgarian Airlines Balkan on 1 April, 1968

Il-14

Air Mali Bamako–Conakry–Monrovia, Bamako–Mopti–Ouagadougou–Accra, Bamako–Nioro–Kayes–Bamako and Bamako–Segou–Mopti–Goundam–Tombouctou–Gao
CAAC (China) Canton–Changkiang–Haikow, Canton–Nanning–Kunming, Chengtu–Chungking–Kunming, Chengtu–Sichang, Lanchow–Chiuchuan–Hami–Urumchi, Peking–Chengchow–Wuhan–Changsha–Canton, Peking–Chengchow–Wuhan–Kweiyang–Kunming, Peking–Hanoi, Peking–Nanking–Shanghai, Peking–Paotow–Yinchuan–Lanchow, Peking–Pyongyang, Peking–Taiyuan–Sian–Chengtu, Peking–Taiyuan–Sian–Lanchow, Peking–Tsinan–Hofei–Shanghai, Peking–Tsinan–Nanking–Shanghai, Peking–Shenyang–Changchun–Harbin, Peking–Shenyang–Harbin, Peking–Sian–Chungking–Kunming, Peking–Wuhan–Nanning, Shanghai–Hangchow–Nanchang–Canton, Shanghai–Hofei–Wuhan–Chungking–Chengtu, Shanghai–Nanking–Changsha–Kweiyang–Kunming, Shanghai–Nanking–Chengchow–Sian–Lanchow, Shanghai–Nanking–Wuhan–Chungking–Chengtu, Shanghai–Wuhan–Chungking, Shenyang–Peking–Taiyuan–Sian, Shenyang–Peking–Taiyuan–Sian–Chengtu, Shenyang–Tientsin–Tsinan–Nanking–Shanghai and Sian–Yenan–Yulin
Cubana Havana–Cienfuegos–Camaguey–Santiago and Havana–Nueva Gerona
ČSA Košice–Bratislava, Košice–Ostrava–Brno, Olomouc/Přerov–Brno–Prague, Prague–Bratislava, Prague–Bratislava–Košice, Prague–Bratislava–Sliač (Banská Bystrica), Prague–Bratislava–Sliač–Košice, Prague–Brno, Prague–Brno–Bratislava–Sliač, Prague–Brno–Gottwaldov (Holešov), Prague–Brno–Ostrava, Prague–Gottwaldov, Prague–Gottwaldov–Ostrava, Prague–Karlovy Vary, Prague–Košice, Prague–Olomouc/Přerov–Košice, Prague–Olomouc/Přerov–Ostrava, Prague–Ostrava, Prague–Ostrava–Košice, Prague–Piešťany, Prague–Piešťany–Sliač, Prague–Sliač, Prague–Sliač–Košice and Prague–Vienna
Malév Budapest–Belgrade–Tirana, Budapest–Berlin, Budapest–Bucharest, Budapest–Debrecen, Budapest–Dubrovnik, Budapest–Pecs, Budapest–Prague, Budapest–Sofia, Budapest–Szombathely–Zalaegerszeg, Budapest–Vienna–Berlin and Budapest–Zagreb–Dubrovnik
Polskie Linie Lotnicze (LOT) Katowice–Gdańsk, Warsaw–Gdańsk, Warsaw–Katowice, Warsaw–Koszalin, Warsaw–Kraków, Warsaw–Poznań, Warsaw–Rzeszów, Warsaw–Wrocław and Wrocław–Gdańsk
Tabso Sofia–Burgas, Sofia–Burgas–Varna, Sofia–Gorna Oryakhovitsa/Tŭrnovo, Sofia–Khaskovo–Burgas, Sofia–Plovdiv–Burgas–Varna, Sofia–Stara Zagora, Sofia–Varna and Varna–Burgas
Tarom Bucharest–Bacău–Suceava, Bucharest–Cluj, Bucharest–Cluj–Baia-Mare, Bucharest–Constanta, Bucharest–Craiova–Timişoara and Bucharest–Oradea Mare

Il-18

Air Mali Bamako–Abidjan, Bamako–Casablanca–Paris, Bamako–Dakar and Bamako–Monrovia–Abidjan–Accra–Douala–Brazzaville
CAAC (China) Canton–Hangchow–Peking, Peking–Chengtu–Kunming, Peking–Irkutsk, Peking–Rangoon and Peking–Shanghai–Canton
ČSA Bratislava–Sofia, Košice–Bratislava, Prague–Amsterdam–Brussels, Prague–Ankara–Beirut–Damascus, Prague–Bratislava, Prague–Brno, Prague–Bucharest–Damascus–Baghdad, Prague–Budapest–Sofia, Prague–Copenhagen, Prague–Dubrovnik, Prague–Kiev, Prague–Košice, Prague–London, Prague–Ostrava, Prague–Rabat–Dakar–/Bamako/–Conakry, Prague–Rabat–Dakar–Freetown, Prague–Sofia, Prague–Split, Prague–Warsaw and Prague–Zagreb–Belgrade. By September Il-18Ds were working Prague–Shannon–Gander–Havana services

Il-18 cont.
Cubana Havana–Santiago
Interflug Berlin–Beirut, Berlin–Belgrade–Dubrovnik, Berlin–Bucharest, Berlin–Budapest, Berlin–Budapest–Algiers–Bamako–Conakry, Berlin–Budapest–Tirana, Berlin–Cairo, Berlin–Kiev, Berlin–Moscow, Berlin–Nicosia–Damascus–Baghdad, Berlin–Prague, Berlin–Sofia and Berlin–Zagreb–Belgrade
Malév Budapest–Athens–Cairo, Budapest–Athens–Damascus, Budapest–Athens–Nicosia, Budapest–Beirut, Budapest–Belgrade–Nicosia, Budapest–Berlin, Budapest–Berlin–Copenhagen–Stockholm–Helsinki, Budapest–Brussels, Budapest–Brussels–London, Budapest–Frankfurt-am-Main–Paris, Budapest–Istanbul–Cairo, Budapest–Kiev, Budapest–London, Budapest–Milan, Budapest–Moscow, Budapest–Munich, Budapest–Paris, Budapest–Prague–Amsterdam, Budapest–Rome, Budapest–Sofia, Budapest–Warsaw, Budapest–Zürich, Budapest–Zürich–Brussels and Sofia–Budapest–Berlin
Polskie Linie Lotnicze (LOT) Warsaw–Amsterdam–Brussels, Warsaw–Beirut, Warsaw–Berlin–Amsterdam, Warsaw–Berlin–Brussels, Warsaw–Berlin–London, Warsaw–Berlin–Paris, Warsaw–Bucharest, Warsaw–Bucharest–Beirut, Warsaw–Budapest, Warsaw–Copenhagen, Warsaw–London, Warsaw–Milan, Warsaw–Moscow, Warsaw–Paris, Warsaw–Rome, Warsaw–Shannon, Warsaw–Sofia, Warsaw–Split, Warsaw–Vienna–Athens–Cairo, Warsaw–Zagreb, Warsaw–Zürich and Warsaw–Zürich–Paris
Tabso Sofia–Athens–Damascus, Sofia–Athens–Tunis–Algiers, Sofia–Bucharest–Moscow, Sofia–Budapest–Berlin–Copenhagen, Sofia–Istanbul–Beirut, Sofia–Moscow, Sofia–Nicosia, Sofia–Paris–London, Sofia–Prague–Berlin, Sofia–Varna, Sofia–Vienna–Frankfurt-am-Main, Sofia–Vienna–Paris, Sofia–Vienna–Paris–London and Sofia–Zürich–London
Tarom Bucharest–Arad–Timişoara, Bucharest–Brussels–London, Bucharest–Budapest–Berlin, Bucharest–Budapest–Prague, Bucharest–Budapest–Warsaw, Bucharest–Frankfurt-am-Main, Bucharest–Istanbul–Beirut, Bucharest–Moscow, Bucharest–Paris, Bucharest–Prague–Berlin–Copenhagen, Bucharest–Prague–Copenhagen, Bucharest–Rome, Bucharest–Sofia–Athens, Bucharest–Sofia–Athens–Cairo, Bucharest–Vienna, Bucharest–Vienna–Paris, Bucharest–Vienna–Zürich and Bucharest–Zürich–Paris

Tu-104A

ČSA Prague–Athens–Cairo, Prague–Athens–Cairo–Dubai–Bombay–Rangoon–Singapore–Djakarta, Prague–Beirut–Bahrein–Bombay–Rangoon–Pnom-Penh, Prague–Berlin–Stockholm–Helsinki, Prague–London, Prague–Milan, Prague–Moscow, Prague–Paris, Prague–Rome, Prague–Tunis–Algiers and Prague–Zürich

Tu-114

Japan Air Lines Tokyo–Moscow (in association with Aeroflot)

Tu-124

ČSA Prague–Amsterdam, Prague–Berlin–Copenhagen, Prague–Berlin–Stockholm–Helsinki, Prague–Bratislava, Prague–Brussels, Prague–Budapest–Bucharest, Prague–Budapest–Sofia, Prague–Copenhagen, Prague–Frankfurt-am-Main, Prague–Košice, Prague–Paris, Prague–Vienna, Prague–Warsaw and Prague–Zürich

APPENDIX V

World Records held by Soviet Transport Aircraft—1967

(Officially homologated by Fédération Aéronautique Internationale)

Class C—landplanes, seaplanes and amphibians
Group II—propeller-turbine powered
Sub-class C-1—landplanes

General

Speed over 500 km (310 st miles) closed circuit: An-10 (pilot A. Mitronine) 29 April, 1961, 730·610 km/h (453·979 mph)
Speed over 1,000 km (621 st miles) closed circuit: Tu-114 (pilot I. Soukhomline) 24 March, 1960, 871·38 km/h (541·447 mph)
Speed over 2,000 km (1,242 st miles) closed circuit: Tu-114 (pilot I. Soukhomline) 1 April, 1960, 857·277 km/h (532·691 mph)
Speed over 5,000 km (3,106 st miles) closed circuit: Tu-114 (pilot I. Soukhomline) 9 April, 1960, 877·212 km/h (545·07 mph)
Speed over 10,000 km (6,213 st miles) closed circuit: Tu-114 (pilot I. Soukhomline) 21 April, 1962, 737·352 km/h (458·168 mph)

Speed Records with Payload

With 1,000 kg (2,204 lb), 2,000 kg (4,409 lb), 5,000 kg (11,023 lb), 10,000 kg (22,046 lb), 15,000 kg (33,069 lb), 20,000 kg (44,092 lb) and 25,000 kg (55,116 lb)
Speed over 1,000 km (621 st miles): Tu-114 (pilot I. Soukhomline) 24 March, 1960, 871·38 km/h (541·447 mph)
Speed over 2,000 km (1,242 st miles): Tu-114 (pilot I. Soukhomline) 1 April, 1960, 857·277 km/h (532·691 mph)
Speed over 5,000 km (3,106 st miles): Tu-114 (pilot I. Soukhomline) 9 April, 1960, 877·212 km/h (545·07 mph)
With 1,000 kg (2,204 lb), 2,000 kg (4,409 lb), 5,000 kg (11,023 lb) and 10,000 kg (22,046 lb)
Speed over 10,000 km (6,213 st miles): Tu-114 (pilot I. Soukhomline) 21 April, 1962, 737·352 km/h (458·168 mph)

Altitude Records with Payload

10,000 kg (22,046 lb) load: Il-18 (pilot V. Kokkinaki) 15 November, 1958, 13,154 m (42,713 ft)
15,000 kg (33,069 lb) load: Il-18 (pilot V. Kokkinaki) 14 November, 1958, 12,471 m (40,915 ft)
20,000 kg (44,092 lb) load: Il-18 (pilot V. Kokkinaki) 25 November, 1959, 12,118 m (39,757 ft)
25,000 kg (55,116 lb) load: Tu-114 (pilot I. Soukhomline) 12 July, 1961, 12,073 m (39,609 ft)
30,000 kg (66,139 lb) load: Tu-114 (pilot I. Soukhomline) 12 July, 1961, 12,073 m (39,609 ft)
35,000–100,000 kg (77,162–220,462 lb) loads in increments of 5,000 kg (11,023 lb): An-22 (pilot I. Davydov) 26 October, 1967, 7,800 m (25,590 ft)

Greatest Load Carried to 2,000 m (6,561 ft)

An-22 (pilot I. Davydov) 26 October, 1967, 100,444 kg (221,440 lb)

Class C—Group II—Sub-class C-2 seaplanes

General

Altitude: M-10 (Be-10) (pilot G. Bourianov) 9 September, 1961, 14,962 m (49,088 ft)
Speed over 15–25 km (9–15 st miles): M-10 (Be-10) (pilot N. Andrievsky) 7 August, 1961, 912 km/h (566·68 mph)
Speed over 1,000 km (621 st miles) closed circuit: M-10 (Be-10) (pilot G. Bourianov) 3 September, 1961, 875·86 km/h (544·237 mph)

Speed Records with Payload

With 1,000 kg (2,204 lb), 2,000 kg (4,409 lb) and 5,000 kg (11,023 lb)
Speed over 1,000 km (621 st miles) closed circuit: M-10 (Be-10) (pilot G. Bourianov) 3 September, 1961, 875·86 km/h (544·237 mph)

Altitude Records with Payload

1,000 kg (2,204 lb), 2,000 kg (4,409 lb) and 5,000 kg (11,023 lb): M-10 (Be-10) (pilot G. Bourianov) 8 September, 1961, 14,062 m (46,135 ft)
10,000 kg (22,046 lb) load: M-10 (Be-10) (pilot G. Bourianov) 11 September, 1961, 12,733 m (41,774 ft)
15,000 kg (33,069 lb) load: M-10 (Be-10) (pilot G. Bourianov) 12 September, 1961, 11,997 m (39,360 ft)

Greatest Load Carried to 2,000 m (6,561 ft)

M-10 (Be-10) (pilot G. Bourianov) 12 September, 1961, 15,206 kg (33,523 lb)

Class E—(giravions)
Sub-class E-1—helicopters

General

Altitude (women): Mi-4 (pilot Tatiana Roussian) 12 January, 1965, 7,524 m (24,685 ft)
Speed over 100 km (62 st miles) closed circuit: Mi-6 (pilot B. Galitsky) 26 August, 1964, 340·15 km/h (211·36 mph)

Speed over 500 km (310 st miles) closed circuit: Mi-6 (pilot B. Galitsky) 15 September, 1962, 315·657 km/h (196·138 mph)

Speed over 1,000 km (621 st miles) closed circuit: Mi-6 (pilot B. Galitsky) 15 September, 1962, 300·377 km/h (186·643 mph)

Speed over 100 km (62 st miles) closed circuit (women): Mi-2 (pilot Tatiana Roussian) 20 June, 1965, 269·38 km/h (167·381 mph)

Speed over 500 km (310 st miles) closed circuit (women): Mi-4 (pilot S. Kotova) 21 July, 1965, 195·447 km/h (121·444 mph)

Speed Records with Payload

With 1,000 kg (2,204 lb) and 2,000 kg (4,409 lb)

Speed over 1,000 km (621 st miles) closed circuit: Mi-6 (pilot B. Galitsky) 15 September, 1962, 300·377 km/h (186·643 mph)

With 5,000 kg (11,023 lb)

Speed over 1,000 km (621 st miles) closed circuit: Mi-6 (pilot V. Kolochenko) 11 September, 1962, 284·354 km/h (176·684 mph)

Altitude Records with Payload

5,000 kg (11,023 lb) load: Mi-10K (pilot V. Kolochenko) 26 May, 1965, 7,151 m (23,461 ft)

10,000 kg (22,046 lb) load: Mi-6 (pilot G. Karapetyan) 16 April, 1959, 4,885 m (16,026 ft)

15,000 kg (33,069 lb), 20,000 kg (44,092 lb) and 25,000 kg (55,116 lb) loads: Mi-10K (pilot G. Alferov) 28 May, 1965, 2,840 m (9,317 ft)

Greatest Load Carried to 2,000 m (6,561 ft)

Mi-6 (pilot G. Karapetyan) 13 September, 1962, 20,117 kg (44,350 lb)

APPENDIX VI

Airline Fleets

It is obviously not possible to give complete production lists of Soviet aircraft; however, enough individual Soviet-designed aircraft are known to make their listing worthwhile.

In many instances the non-Soviet fleets are complete.

All Soviet-registered aircraft have been listed under Aeroflot, although in fact a few of these may be registered to other organizations or to design bureaux.

An-2

Aeroflot (all with SSSR- prefix)					
A462				05902	An-2M
A4163		An-2S		05905	An-2M
L462	10847303	An-2S		05908	An-2M
L1925		An-2S		05912	An-2M
L1938				05926	An-2M
L1939				06231	
L1967				07955	
L1972				07960	
L1978				09601	An-2S
L2588	10447361			23667	
L2783				23708	An-2P
L3470				23749	An-2M
L3733		An-2P*		23775	
L3749		An-2S		23812	An-2V
L3753		An-2PS		25454	
L3814				25474	
L3815		An-2V		25478	
L5483				25584	An-2P
L5516	14147313	An-2P		28930	An-2S*
L5584	15047303	An-2S		28932	
L5612				32610	An-2S
L5614				33159	An-2M
L5878				33177	An-2S
N542			Polar	35422	
01205				42641	An-2M
01275			Polar	42643	An-2M
02162		An-2P floatplane		47643	An-2P
04275				62713	
04351			Polar	62729	An-2M
05899				62760	
05901		An-2M		70880	
				70888	

* Unconfirmed

296

70988			93480		
79863			98261		An-2P
79954			98288		
79977			98302		
91504			98340		

Air Mali

TZ-ABS	116202		TZ-ABT	116204

Deutsche Lufthansa (DDR) and Interflug

DM-SKA	An-2P	DLH, to Interflug		DM-SKF	An-2P	DLH, to Interflug	
DM-SKB	,,	,,	,,	DM-SKG	,,	,,	,,
DM-SKC	,,	,,	,,	DM-SKI	,,	,,	,,
DM-SKD	,,	,,	,,	DM-SKK	,,	,,	,,
DM-SKE	,,	,,	,,	DM-SKL	,,	,,	,,

Royal Nepal Airlines Corporation

9N-AAJ*   9N-AAK*

An-10 and 10A

Aeroflot (all with SSSR- prefix)

U1957		An-10 prototype	11191		An-10A
11129	0402506		11195		
11134			11196		
11135		An-10	11203		
11136			11205		
11137			11217		
11138			11219		An-10A
11140		An-10A	11220	0402801	An-10A
11141		An-10	11221	0402102	
11144			11222		
11146		An-10	11229		
11148		An-10A	34385		
11156					
11157		An-10			
11158		An-10			
11159		An-10			
11160					
11161					
11170		An-10A			
11172	9401602	An-10A			
11175		An-10A			
11179					
11185		An-10A			
11188					

* Built in China and known as Fongshu or Harvester 2.

An-12

Aeroflot (all with SSSR- prefix)

04343		Polar	11367	402901	
04364		Polar	11373	402403	
04366		Polar	11381	402807	With skis
11031	7345003		11385		
11340			11387		
11341	401702		11397		
11353			33688		
11357		Polar	72613		
11359	402804	An-12B	75622		
11361	402704		75625		
11366	402808				

Cubana

CU-T827

Ghana Airways

9G-AAZ 024009

An-14 Pchelka

Aeroflot (all with SSSR- prefix)

L1053	L5860	
L1956	81550	Production aircraft
L1958		

Tabso

LZ-TED

An-24

Aeroflot (all with SSSR- prefix)

L1959		First prototype	46718		
L1960		Second prototype	46719		
L19605		An-24V	46720		
46202		An-24V	46721		
46217		An-24V	46722		
46248		An-24V	46723		
46280	7910104	An-24TV	46724		
46708			46727		
46709			46730		
46710			46731		
46711			46735		
46714			46736		
46715			46740		
46716			46756		An-24V
46717			46763		An-24V

46776		An-24V	46806			An-24V
46785			46813			An-24V
46791	57301804	An-24V	46848			An-24V
46803		An-24V	98104		67302310	An-24RV

Air Mongol - Mongolian Airlines

HMAY-1202

Cubana (An-24V)

CU-T875		CU-T879
CU-T876		CU-T880
CU-T877		CU-T881
CU-T878		CU-T882

Interflug (An-24V)

DM-SBA		DM-SBE	02301
DM-SBB		DM-SBF	
DM-SBC	02302	DM-SBG	02303
DM-SBD	02210	DM-SBH	

Lebanese Air Transport

OD-AEN   to Misrair SU-AOM

Misrair* (An-24V)

SU-ANV	01805	SU-AOB		
SU-ANW		SU-AOC	02103	
SU-ANX	02002	SU-AOK	02805	
SU-ANY	02004	SU-AOL	02806	
SU-ANZ	02007	SU-AOM		ex OD-AEN
SU-AOA				

Polskie Linie Lotnicze (LOT)

SP-LTA	02203		SP-LTG	02504
SP-LTB	02205		SP-LTH	02505
SP-LTC	02208		SP-LTI	02506
SP-LTD	02209		SP-LTK	02507
SP-LTE	02405		SP-LTL	
SP-LTF	02406			

Tabso

LZ-ANA	02707		LZ-ANE	
LZ-ANB	02710		LZ-ANF	03407
LZ-ANC			LZ-ANG	03408
LZ-AND				

* Now merged with United Arab Airlines

An-24 *cont*.
Tarom

YR-AMX	02107	YR-AMZ	
YR-AMY			

Il-12

Aeroflot (all with SSSR- prefix)

L1037		L1766		
L1302		L1768		
L1306		L1771		
L1308		L1783		
L1312		L1786		
L1314		L1789		
L1320		L1790		
L1344		L1794		
L1346		L1796		
L1348		L1802		
L1371		L1805		
L1380		L1806		
L1384		L1809		
L1390		L1810		
L1397		L1812		
L1401		L1815		
L1403		L1816		Il-12B
L1413		L1817		
L1416		L1819	13319	
L1418		L1822		Il-12B
L1445		L1823		
L1458	cargo	L1824		
L1461		L1826		
L1638	Il-12B	L1827		
L1701		L1828		
L1702		L1829		
L1704		L1830		
L1719		L1832		
L1720		L1833		
L1721		L1834		
L1723	Il-12B	L1835		
L1724		L1837		
L1741		L1838		
L1742		L1840		
L1743		L1842		
L1744		L1851		
L1750		L1860		
L1754		L1866		
L1756		L1894		
L1757		L1921		
L1761		L1926		

L1934			N440		Il-12B Polar	
L4155			N479			Polar
N438		Polar	01440		Il-12B	

ČSA (Il-12B)

OK-CBA	83012904	OK-DBB	
OK-CBB		OK-DBC	
OK-CBC		OK-DBD	
OK-CBD		OK-DBG*	
OK-CBE		OK-DBO	
OK-CBF		OK-DBP	
OK-CBG*		OK-DBU*	
OK-CBH		OK-DBW*	
OK-CBJ		OK-DNB*	
OK-CBK			

Polskie Linie Lotnicze (LOT) (Il-12B)

SP-LHA	01		SP-LHD	04
SP-LHB	02		SP-LHE	05
SP-LHC	06			

Il-14

Aeroflot (all with SSSR- prefix)

L1324†			L1626	146000927	
L1334	147001334		L1628	146000929	
L1358	147001430		L1632	146000930	
L1386	147001449		L1633	146000933	
L1435†	147001519	to SSSR-61778	L1634	146000934	
L1465			L1636	146000936	
L1502	146001125		L1638	146001020	
L1510	147001329		L1657		
L1515			L1658	146001028	
L1526	147001407		L1673		
L1531			L1676	146001039	
L1533	147001149		L1688	146001113	Il-14M
L1534			L1697		
L1537			L1729		Il-14P
L1548	147001212		L1803	146000938	
L1552	147001411		L1818	146000810	
L1555		Il-14M	L1819		
L1563	147001303		L1824	434035	Il-14P
L1566			L1834		
L1572			L1836		
L1576	147001315		L1844		Il-14P
L1580	147001309		L1849		
L1605	146000844		L1866		
L1625	146000925		L1869	146000840	Il-14P

* Converted to cargo aircraft.
† Cargo aircraft.

Il-14 cont.

L1870	146000401	Il-14P	41897		
L1872		Il-14P	52000		
L1873			52008		
L1874	146000607	Il-14P	52019		
L1875	146000609	Il-14P	52029		
L1876			52037		
L1877	146000613		52042		
L1878	146000615		52043		
L1880	146000841		52051		
L1886			52052		
L1889	146000801		52064		
L1890	146000714	Il-14P	52065		
L1891	146000715	Il-14P	52075		
L1895			52085		
L1896			52094		
L2040			52671		
L5007			61622		
L5011			61671		
L5040		Il-14M	61704		
L5047		Il-14M	61722		Il-14M
L5051			61725		
L5058			61757		
L5060	4340304	Il-14P	61778*	147001519	ex SSSR- L1435
L5072			61790		
L5075			61796		
L5076			61797		
L5081			61798		
L5085			61800		
L5086		Il-14M	91481		
L5088			91483		
01404			91490	147001501	
01450			91491		
04177		Polar	91514		Il-14M
04179		Polar	91520		
33682			91542		
41817			91551		
41821			91557	147001235	Il-14M
41848	6341706		91578		
41882			91582		
41889			91584		

Air Guinée

3X-BKE	013173	Avia-14T	3X-PRG			
3X-CKY	014146	Avia-14 Super	3X-SIG†		Avia-14-32A	
3X-KDA		Avia-14-32A	OK-MCO	806108	Avia-14-32A‡	
3X-PDG						

* Cargo aircraft.　　† Reregistered 3X-GAH　　‡ Leased from ČSA.

Air Mali

TZ-ABF	147001310	Avia-14-32A	TZ-ABH	7342501	Avia-14-32A
TZ-ABG	146001050	Avia-14-32A			

Air Mongol - Mongolian Airlines
HMAY-101

CAAC (China)

606	Il-14P	654	Il-14M
614		660	
618		672	Il-14P
620	Il-14P	4214	
634			

ČSA

OK-LCA	703109	Avia-14P	converted to cargo aircraft
OK-LCB	703110	,,	,, ,,
OK-LCC	703111	,,	,, ,,
OK-LCD	703112	,,	,, ,,
OK-LCE	703113	,,	,, ,,
OK-LCF	705108	,,	,, ,,
OK-MCG	805109	Avia-14-32A	
OK-MCH	805118	,,	Písek
OK-MCI	805119	,,	Vlaštovka
OK-MCJ	805120	,,	Gottwaldov
OK-MCK	806104	,,	Strakonice
OK-MCL	806105	,,	Trenčín
OK-MCM	806106	,,	
OK-MCN	806107	,,	
OK-MCO	806108	,,	some time leased to Air Guinée
OK-MCP	806109	,,	
OK-MCR	806110	,,	
OK-MCS	806111	,,	
OK-MCT		,,	
OK-MCU	807104	,,	Bílá Labuť
OK-MCV	807106	,,	Lučenec
OK-MCW	807107	,,	Olomouc
OK-MCX	807108	,,	Přerov
OK-MCY	807109	,,	Jindřichův Hradec
OK-MCZ	807110	,,	
OK-OCA	013167	Avia-14T	

Cubana

CU-T814	CU-T818	CU-T822
CU-T815	CU-T819	CU-T823
CU-T816	CU-T820	CU-T824
CU-T817	CU-T821	CU-T825

Il-14 *cont.*
Deutsche Lufthansa (DDR) and Interflug

DDR-ABA	5340709	Il-14P	DLH*†
DDR-AVF		VEB Il-14P	DLH
DM-SAA		,,	Interflug
DM-SAB		,,	,,
DM-SAC		,,	,,
DM-SAD		,,	,,
DM-SAE		,,	,,
DM-SAF		,,	,,
DM-SAG		,,	,,
DM-SAH	14803020	,,	DLH†
DM-SAI	14803023	,,	,, †
DM-SAK	14803019	,,	,, †
DM-SAL	14803026	,,	,, †
DM-SAM		,,	,,
DM-SAN		,,	,,
DM-SAO	14803079	,,	Interflug
DM-SAR		,,	,,
DM-SBA	5340709	Il-14P	,, ‡
DM-SBB		,,	,,
DM-SBC		,,	,,
DM-SBD		,,	,,
DM-SBE		,,	,,
DM-SBF		,,	,,
DM-SBG		,,	,,
DM-SBH		,,	,,
DM-SBI		,,	,,
DM-SBK		,,	,,
DM-SBL		,,	,,
DM-SBM		,,	,,
DM-SBN		,,	,,
DM-SBO		,,	,,
DM-SBP		,,	,,
DM-SBR		,,	,,
DM-SBS		,,	,,
DM-SBU		,,	,,
DM-SBV		,,	,,
DM-SBW		,,	,,
DM-SBX		,,	,,
DM-SBY		,,	,,
DM-SBZ		,,	,,

Jugoslovenski Aerotransport (JAT)

YU-ADE			YU-ADH	147001341	Il-14P
YU-ADF			YU-ADI		
YU-ADG			YU-ADJ		

* Reregistered DM-SBA. † To Interflug. ‡ Ex DDR-ABA.

Malév

HA-MAA	147001325	Il-14M		HA-MAG		VEB Il-14P
HA-MAB	147001423	,,		HA-MAH	14803033	,,
HA-MAC	147001424	,,		HA-MAI	14803034*	,,
HA-MAD	14803028	VEB Il-14P		HA-MAK	6341102	
HA-MAE	14803029	,,		HA-MAN		

Polskie Linie Lotnicze (LOT)

SP-LNA	4340607	Il-14P		SP-LNI	14803013	VEB Il-14P
SP-LNB	4340510	,,		SP-LNK	14803014	,,
SP-LNC	4340509	,,		SP-LNL	14803017	,,
SP-LND	6341404	,,		SP-LNM	14803018	,,
SP-LNE	6341602	,,		SP-LNN	703108	Avia-14
SP-LNF	6341407	,,		SP-LNO	14803065	VEB Il-14P
SP-LNG	14803010	VEB Il-14P		SP-LNP	14803069	,,
SP-LNH	14803012	,,		SP-LNR	14803055	,,

Tabso

LZ-ILA	146001046	Il-14P		LZ-ILG		Avia-14 Super
LZ-ILB	146001103	Il-14M		LZ-ILK		
LZ-ILC	14803038	VEB Il-14P		LZ-ILM		
LZ-ILD	14803039	,,		LZ-ILO		
LZ-ILE	14803040	,,		LZ-ILP		
LZ-ILF		Avia-14 Super				

Tarom

YR-ILA	146000926	Il-14P		YR-ILJ	14803070	VEB Il-14P
YR-ILB	146000902	,,		YR-ILK	14803068	,,
YR-ILC	146000924	,,		YR-ILL	14803072	,,
YR-ILD	146001036	,,		YR-ILM	14803074	,,
YR-ILE	146001117	Il-14M		YR-ILN	14803060	,,
YR-ILF	146001116	,,		YR-ILO	14803061	,,
YR-ILG	146001119	,,		YR-ILP	14803062	,,
YR-ILH	14803064	VEB Il-14P		YR-ILX		
YR-ILI	14803066	,,		YR-ILZ	148002001	Il-14P

United Arab Airlines

SU-ANE	Il-14P

Yemen Airlines

YE-AAG	Il-14P

* c/n may be 14800034.

Il-18

Aeroflot (all with SSSR- prefix)

L5811			75546		
L5818			75547	184007302	
L5819			75552		
L5820			75553	184007405	
L5821			75554		
04330		Polar	75558		
04350		Polar	75559		
04770		Polar	75560	184007704	
33569			75562		
75400	186008901		75563		
75401	186008902	Il-18D	75568		
75402	186008903		75569		
75403			75570		
75404	186009003		75581	185007803	Il-18D
75412	186009301	Il-18D	75586	185008401	
75413	186009302		75587	185008402	
75414			75588		
75418			75590		
75419			75591		
75430		Il-18D	75595		
75432	186009501		75601		
75435	186009504	Il-18D	75606		
75436			75607		
75438			75631		
75441		Il-18D	75633		
75444	187009801		75639		
75446			75644		
75448	187010003		75645		
75449	187010004		75648		
75452	187010102		75649		
75487			75650		
75502			75651		
75504			75653		
75508			75654		
75512	183006505		75655		
75515			75656		
75518			75661		
75520			75662		
75523			75663		
75525	183006803		75664		
75528			75666		
75530	184006903		75668		
75533			75669		
75535			75672		
75538	184007002		75673		
75539			75674		
75540			75675		
75545	184007204		75676		

75677			75749		
75678	1003		75753		
75679			75756		
75680			75759		
75681			75760		
75683			75761		
75684			75763		
75686			75764	181003402	to 7T-VRA
75687			75765		
75688			75766	181003405	
75693			75768	189001201	
75694			75769		
75695			75770		
75701			75771	181003505	
75702	189001503		75772	181003601	
75703			75773	181003603	
75704			75774	181003604	
75705	189001702		75775	181003605	
75706	189001703		75776	181003701	
75707	189001704		75780		
75708			75784		
75710			75786		
75711			75788	181004002	Il-18V
75712			75789	181004003	
75713			75790	181004004	
75714			75791	181004005	
75715			75792	181004101	
75716	180001902		75793	181004102	
75717			75794	181004103	
75718	180002003		75795	181004104	
75719	180002004		75797		
75720			75799	181004204	
75721	180002105		75803	182004304	
75722	180002201		75804		
75723		Il-18V	75812		
75724			75813		
75725			75816	182004702	
75726			75817	182004703	
75727			75818	182004704	
75728			75820	182004805	
75730	180002401		75821		
75731			75822		
75733			75823	182004902	Il-18V
75736			75824	182004903	Il-18V
75737	181002702		75825	182004904	
75739	181002704		75828	182005002	
75740	181002705		75835		
75742	181002802	Il-18V	75841		Il-18V leased to
75743	181002901				Air Guinée
75745					
75748			75842		
			75845	182005305	

Il-18 *cont.*

75848		75872	
75851		75873	183006003
75852		75874	183006004
75860		75878	
75862		75879	
75863		75880	183006103
75865		75882	183006105
75867		75888	Il-18I
75869		75894	
75870			

Air Guinée

3X-KKN	181003703	to 3X-GAB	3X-NZE	180002004	ex SSSR-75719
3X-LBE	181003705	to 3X-GAA	3X-NZE^2	181003704	to 3X-GAC

Air Mali

TZ-ABD	181003303		TZ-ABY	182004505
TZ-ABE	181003304			

CAAC (China)

202		238*
206		502*

ČSA

OK-NAA	189001604	Il-18	*Ostrava*, later *Piešťany*
OK-NAB	189001605	Il-18	*Košice*
OK-OAC	180002101	Il-18V	*Sliač*
OK-OAD	180002102	Il-18V	
OK-PAE	181002902	Il-18V	*Karlovy Vary*
OK-PAF	181002904	Il-18V	
OK-PAG	181004201	Il-18V	*Vysoké Tatry*
OK-PAH	181004202	Il-18V	*Mariánské Lázně*
OK-WAI	187009705	Il-18D	
OK-WAJ	187010101	Il-18D	

Cubana

CU-T830	
CU-T831	182005202
CU-T832	

* Unconfirmed, also reported as 204 and 208.

Deutsche Lufthansa (DDR) and Interflug

DM-STA	180001905	DLH, to Interflug	DM-STE	182005101	Interflug
DM-STB	180002001	DLH, to Interflug	DM-STF	181004105	,,
			DM-STG	182004402	,,
DM-STC	180002202	DLH, to Interflug	DM-STH	184007305	,,
			DM-STI		,,
DM-STD		Interflug	DM-STK	186009202	,,
			DM-STL		,,

Ghana Airways (Il-18V)

9G-AAI	180002402		9G-AAM	181003305
9G-AAJ	2405		9G-AAN	181003403
9G-AAK	2501		9G-AAX	181002803
9G-AAL	2502		9G-AAY	181002804

Malév (Il-18V)

HA-MOA	180001903		HA-MOF	183006301
HA-MOC	181002903		HA-MOG	184007103
HA-MOD	180002002		HA-MOH	184007104
HA-MOE	182005505		HA-MOI	

Polskie Linie Lotnicze (LOT)

SP-LSA	180002403	Il-18V	SP-LSE	183006504	Il-18V
SP-LSB	180002404	Il-18V	SP-LSF	8601	Il-18V
SP-LSC	181002805	Il-18V	SP-LSG	8603	Il-18V
SP-LSD	184007102	Il-18V	SP-LSH	8701	Il-18D

Tabso

LZ-BED	186009002		LZ-BEO	
LZ-BEG	186009101		LZ-BEP	185008105
LZ-BEK	182004603		LZ-BER	184007203
LZ-BEL	182004601		LZ-BES	185008104
LZ-BEM	183005602		LZ-BET	186008904
LZ-BEN	184007101		LZ-BEV	185008201

Tarom

YR-IMA	181003602		YR-IMH	185008301
YR-IMB	181003702		YR-IMI	185008302
YR-IMC	182004802		YR-IMJ	9102
YR-IMD	182004804		YR-IMK	9104
YR-IME	183006205		YR-IML	187009903
YR-IMF	184007105		YR-IMM	187009904
YR-IMG	184007301		YR-IMZ	8602

Il-62

Aeroflot (all with SSSR- prefix)

06156	86663
06176	86664
06300	86665
86661	86666
86662	86671

Mi-4

Aeroflot (all with SSSR- prefix)

L38		31410	
L47		31416	Mi-4P
L54		31417	Mi-4P
L66		31418	Mi-4P
L69	Mi-4S	31419	Mi-4P
L80		31420	Mi-4P
L87		31429	
L89		31479	
L99		31499	
L515		31524	
L0536		31537	Mi-4P
L0542		31540	Mi-4S to Austria
L0547		31571	Mi-4P
L0558		31598	
N86		35244	
N87		35273	
04332		35277	Mi-4P
19120		35278	Mi-4P
19125	Mi-4P	35294	
19128	Mi-4P	35706	
19129	Mi-4P	36513	
19132	Mi-4P	38249	
19133	Mi-4P	38548	
19138	Mi-4P	48973	
19139	Mi-4P	66831	
19145	Mi-4P	66849	
19159		66860	Mi-4P
19178		66870	
19190		66877	
19232	Mi-4P	66881	
28964		66882	Mi-4P
28993		66885	
29001	Mi-4P	66893	
29027	Mi-4P	66895	Mi-4P
29042	Mi-4P	66899	
29049		66914	Mi-4P
29079	Mi-4P	66915	Mi-4P
30524		66935	
31405			

Royal Nepal Airlines Corporation
9N-HAA   9N-HAB

Tu-104

Aeroflot (all with SSSR- prefix)

L5400	Tu-104 prototype	L5460*	Tu-104B*
L5402	Tu-104	42307*	
L5406	,,	42313	
L5412	,, reregis-	42318	Tu-104 ex
	tered		L5412
	42318	42319	Tu-104A
L5413	,,	42326	,,
L5414	,,	42327	
L5415	,,	42334	Tu-104A
L5416	,,	42335	,,
L5417	,,	42336	
L5418	,,	42337	
L5419	,,	42338	
L5420	,,	42341	Tu-104A
L5421	Tu-104A	42342	
L5422		42343	
L5423	Tu-104A	42344	
L5425		42345	Tu-104A
L5427		42347	,,
L5428	Tu-104A	42348	
L5429	,,	42350	
L5430	,,	42353	
L5432	,,	42354	
L5433		42356	
L5434		42357	Tu-104A
L5435		42358	
L5436	Tu-104A	42359	Tu-104A
L5437		42360	,,
L5438	Tu-104A	42361	
L5439		42362	
L5440	Tu-104A	42367	Tu-104A
L5441	,,	42368	,,
L5442	,,	42369	,,
L5443	,,	42370	,,
L5444	,,	42372	,,
L5445	,,	42374	,,
L5446	,,	42375	
L5447	,,	42377	
L5453	,,	42379	Tu-104A
L5458	,,	42380	
L5459		42381	

* Unconfirmed.

Tu-104 *cont.*

42382		Tu-104A
42383		,,
42384	8350604	,,
42385		,,
42386		
42388		Tu-104A
42389	8350704	,,
42390	8350705	,,
42391		,,
42392		
42393	9350803	Tu-104A
42394		,,
42395		,,
42396		
42397	9350902	Tu-104A
42398	9350903	,,
42399		Tu-104B
42400		,,
42401		,,
42402		,,
42403		,,
42404		,,
42405		,,
42407		,,
42408		,,
42409		,,
42410		,,
42411		,,
42412		,,
42415		,,
42416		,,
42417		,,
42418		,,
42419		,,
42420		,,
42422		,,
42423		,,
42424		,,
42425		,,
42426		
42427		Tu-104B
42428		,,
42429		,,
42430		,,
42431		,,
42433		,,
42434		,,
42435		
42436		
42437		
42438		Tu-104B
42439		,,
42440		,,
42442		,,
42444		,,
42446		,,
42448		
42451		Tu-104A
42453		,,
42455		,,
42456	9350905	,,
42457		
42458		
42459	06601901	Tu-104B
42460	06601902	,,
42461	06601903	,,
42462		,,
42463		Tu-104A
42464		,,
42465		Tu-104B
42467		,,
42468		
42469		Tu-104B
42471		,,
42472		,,
42474		,,
42475		,,
42477		,,
42478		
42480		
42482		Tu-104B
42483		,,
42484		,,
42485		,,
42487		,,
42488		
42489		Tu-104B
42490		,,
42491		,,
42492	021605	,,
42493		,,
42495	021703	,,
42496		
42497		Tu-104B
42498		,,
42499		,,
42500		,,
42501		,,
42503		,,
42504		,,

42505			42508	022001	Tu-104B
42507		Tu-104B	42509		,,

Most Aeroflot Tu-104As are believed to have been converted to Tu-104Ds.

ČSA

OK-LDA	76600503	Tu-104A	Praha (*Prague*)
OK-LDB	76600601*	,,	Bratislava
OK-LDC	76600602	,,	Brno
OK-MDE	86601202	,,	Ostrava
OK-NDD	96601803	,,	Plzeň (*Pilsen*)
OK-NDF	9350801	,,	České Budějovice

Tu-114

Aeroflot (all with SSSR- prefix)

L5611		76473
L5700		76474
76459†		76476
76460		76477
76462	Tu-114D	76480
76464‡		76481
76465		76482
76466		76483
76467		76484
76468		76485
76469		76486
76470		76487
76471		76490‡
76472		

Tu-124

Aeroflot (all with SSSR- prefix)

45000	45023
45003	45025
45004	45026
45005	45031
45006	45033
45007	45038
45013	45040
45014	45042
45015	45043
45016	45047
45018	45052
45021	45054
45022	45055

* Unconfirmed.
† Reserved for joint Aeroflot–Japan Air Lines operations.
‡ Used on joint Aeroflot–Japan Air Lines operations.

Tu-124 *cont.*

45056	45082
45059	45083
45063	45089
45067	45090
45069	45091
45072	45092

ČSA

OK-TEA	1503	*Mělník*
OK-TEB	1504	*Centrotex*
OK-UEC	1607	*Mladá Boleslav*

Interflug

DM-SDA   DM-SDB

Tu-134

Aeroflot (all with SSSR- prefix)

45075	65606	
45076	65608	
65600	65610	6350202
65601	65611	
65603		

Yak-12

Aeroflot (all with SSSR- prefix)

L4108	Yak-12M	62613		
L5880	,, 12R	72664		
07870	,, 12M	72806	Yak-12A	
14261	,, 12M	74077	,, 12M	

*   *   *

Late Fleet Additions

Aeroflot	Il-18D	SSSR-75462	
Air Mali	An-24V	TZ-ACT	c/n 04101

INDEX

Administration of Northern Sea Routes, see Glavsevmorput
Aero 45, 279
Aeroflot
 Administration, 22, 31–2
 Agricultural work, 28, 158–9
 Ambulance work, 20, 28
 Cargo, 27–8, 37, 280–1, 283–5, 287–8
 Cold weather operations, 29–30
 Directorates, 22, 31–2, 53–4, 79–81, 96, 99. 103, 106, 116, 121, 137, 140, 154, 159, 165, 171, 179, 208, 226–7, 255, 277–9, 282–9
 Equipment, 30–1, 277–9
 Fleet, 30–1, 296–302, 306–8, 310–14
 History, 20–6
 Main bases, 277–9
 Newspaper carriage, 28, 126
 Polar operation, 29–30
 Route map, 16–17
 Routes, 16–17, 22, 26, 64–5, 80, 88–9, 99, 112, 123–5, 134, 168, 206, 208, 211–12, 220–2, 224, 226, 233, 235, 240, 280–9
 Safety, 32–3
 Size, 30–1
 Traffic, 22–5, 53–4, 93, 198
Agrolet, 45, 154, 162
Air Forces
 Albania, 107
 Czechoslovakia, 107
 Finland, 154
 German Democratic Republic, 116, 127
 Hungary, 45
 India, 56, 58, 107, 162, 166, 227, 232
 Indonesia, 58, 170
 Iraq, 58
 North Vietnam, 170
 Poland, 81, 107, 116
 United Arab Republic, 45, 58, 107, 170
 USSR, 45, 50–1, 57–8, 96, 100–1, 107, 114, 126, 150, 153, 170, 252–3, 258–9
Air France, 28, 59, 61, 64, 280–1
Air Guinée, 43, 81, 104, 106, 116, 290, 302, 308
Air Mali, 43, 81, 89, 106, 116, 290–1, 297, 303, 308, 314
Air Mongol, see Mongolian Airlines
Airline fleets, 296–314

Airports, 33, 36–40
AK-1, vii, 21
ANT-9, 12, 19, 21, 96
ANT-14, 215
ANT-20 *Maxim Gorki*, 215
ANT-20bis, 215
ANT-35, 18, 96
Antei, see Antonov An-22
Antonov
 An-2, 22, 28, 30–1, 41–9, 89, 93, 152, 269, 277–9, 290, 296–7
 An-2L, 42, 44
 An-2M, 42, 46–9, 296
 An-2P, 41–4, 46–7, 49, 95, 296–7
 An-2R, 49
 An-2S, 42–3, 46, 49, 296
 An-2SN, 49
 An-2T, 49
 An-2V, 42, 44–5, 49, 296
 An-2W, 49
 An-2ZA, 42, 45, 49
 An-4, 44, 50
 An-6, 45
 An-8, 50–1, 63
 An-10, 24–6, 30, 50–64, 71, 78, 113, 225, 246, 269, 277–9, 282, 293, 297
 An-10A, 53–8, 63–4, 278–9, 282, 297
 An-10B, 54
 An-12, 27–30, 37, 55–65, 74, 277–9, 280–1, 283, 298
 An-12B, 60–3, 298
 An-14 Pchelka, 30, 48, 65–71, 93, 298
 An-14A, 70
 An-16, 54
 An-22, 27, 71–8, 261, 273, 294
 An-24, 25, 30, 32, 34, 78–89, 269, 277–9, 283, 290, 298–300, 314
 An-24RV, 83–4, 88, 299
 An-24TV, 83, 85–6, 259, 298
 An-24V, 80–9, 277–9, 298–9, 314
Avia-
 14, 104–7, 109, 305
 14P, 104, 106, 303
 14 Salon, 105–6
 14 Super, 105–7, 302, 305
 14T, 104, 106–7, 302–3
 14-32A, 105–6, 109, 111, 302–3
Aviaarktika, 21–2

Balkan, Bulgarian Airlines, see Tabso
Beriev
  Be-6, 89–92
  Be-10, 89–92, 294
  Be-30, 26, 93–6
  Be-R-1, 91
  LL-143, 90–1
Bulgarian Airlines Balkan, see Tabso

CAAC (China), 43, 100, 106, 116, 290–1, 303, 308
Capital No.1 Sha-Tu, 70
Československé Aerolinie (ČSA), 43, 45, 100, 105–7, 110, 115–16, 120, 125, 154, 162, 206, 209–10, 212, 227, 230–1, 233, 235, 291–2, 301–3, 308, 313–14
Consolidated Catalina, see MP-7
Cubana, 58–9, 81, 107, 116, 290–2, 298–9, 303, 308
Cyrillic alphabet, viii

De Havilland 34, 21
Deruluft, 11–12, 20–1
Deutsche Lufthansa (DDR), 43, 46, 106–8, 116, 125, 297, 304, 309
Dobroflot, 21
Dobrolet, 12, 19, 21
Dornier
  Komet, 21
  Merkur, 21
  Wal, 89
Dosaaf, 43, 45, 151
Douglas
  C-47, 27, 96
  DC-3, 18, 21, 96

Engine overhaul life, 34
Engines
  Glushenkov, 141
  Helwan E-300, 58
  Ivchenko AI-14, 66, 68, 137, 139, 148, 253–4, 256, 259, 261
  „ AI-20, 33–4, 51, 56, 113, 118–19, 121
  „ AI-24, 83–4
  „ AI-25, 272
  „ AI-26, 151, 155, 262
  Izotov GTD-350, 156–7
  „ TV-2, 178, 181
  Klimov VK-1, 91, 127
  Kuznetsov NK-2M, 50
  „ NK-4, 51, 113
  „ NK-8, 34, 130, 244, 248
  „ NK-12, 77, 219
  „ NK-144, 35, 241, 244–5
  Lyulka AL-7, 91

Mikulin AM-3, 201, 204
  „ RD-3, 201
  „ RD-3M, 207
Shvetsov ASh-21, 41, 93, 257
  „ ASh-62, 42, 45–6
  „ ASh-72, 90
  „ ASh-73, 91, 112
  „ ASh-82, 97, 101, 160, 166, 257, 262, 265
  „ ASh-90, 195–6
  „ M-11, 14–15, 70, 252, 259
Soloviev D-20P, 231
  „ D-25, 173, 183, 194
  „ D-30, 131, 237
  „ TV-2M, 178
Tumanskii RU-19-300, 83–4
Turbomeca Astazou, 70, 93
TV-2VM, see Soloviev D-25
TVD-10, 93
Vedeneev M-14, 148

Fokker F.III, 11, 20
Fongshu, 43–4, 297

Ghana Airways, 30, 58–9, 116–17, 125, 298, 309
Glavsevmorput, 21–2, 89
GM-1, see Mil Mi-1
Grazhdanskaya Vozduzhnaya Flot, see Aeroflot

Harvester, 43–4, 297
Heinkel He 111, 21
Heliswiss, 36, 170

Il'ya Muromets, 20, 215
Ilyushin
  Il-12, 23–5, 27, 78, 96–110, 112, 198, 257, 269, 300–1
  Il-12B, 24, 29, 97–100, 108, 110, 300–1
  Il-12T, 99
  Il-14, 24–5, 27, 58, 73, 78, 96–111, 163, 198, 225, 269, 277–9, 283–5, 291, 301–5
  Il-14G, 102–3
  Il-14M, 102–3, 106–11, 301–3, 305
  Il-14P, 101–3, 106–8, 110, 301–5
  Il-14T, 102–3, 108, 110
  Il-18 (piston engines), 33, 111–12
  Il-18 (turbines), 24–7, 30, 32–4, 39, 51, 78, 81, 111, 113–25, 225, 246, 269, 277 –81, 285–7, 291–2, 294, 306–9, 314
  Il-18B, 115
  Il-18D, 33, 119–23, 125, 306, 308–9, 314
  Il-18E, 119–22
  Il-18I, 119, 122, 308

Il-18V, 115–19, 121–3, 307–9
Il-20, 126–7
Il-28, 126–7
Il-62, 26, 30–1, 34–5, 85, 128–35, 221–2, 244, 278, 281, 310
Interflug, 43, 46, 81–2, 89, 102, 107, 116, 162, 227, 235, 290, 292, 297, 299, 304, 309, 314

Japan Air Lines, 221–2, 224, 280–1, 292, 313
Jugoslovenski Aerotransport (JAT), 107, 304
Junkers-
  F 13, 21
  Ju 52/3m, 21

Kalinin
  K4, 21
  K5, 12
Kamov
  Ka-8, 136
  Ka-10, 136
  Ka-15, 136–8, 144
  Ka-18, 139–40, 144
  Ka-20, 140
  Ka-25K, 140–4
  Ka-26, 143–9
Kolkhoznik, 41

L-200D Morava, 278–9
Lebanese Air Transport, 80, 82, 299
Lebanese Overseas Airways, 80
Lisunov Li-2, 21, 23, 27–8, 37, 41, 78, 96, 112, 163, 198, 225, 269, 277–9, 287
LL-143, see Beriev LL-143
LOT, see Polskie Linie Lotnicze

M-10, see Beriev Be-10
Malév, 30, 107, 116, 118, 125, 235, 291–2, 305, 309
Maxim Gorki, see ANT-20
Mil
  Mi-1, 28, 35, 150–4, 156, 159–60, 178
  Mi-1A, 152, 154
  Mi-1 Moskvich, 152–3
  Mi-1NKh, 151–2, 154
  Mi-1T, 152, 154
  Mi-1U, 151
  Mi-2, 26, 35–6, 154–60, 167, 178, 277, 295
  Mi-4, 30, 34–5, 159–68, 178–9, 182, 262, 277–9, 294–5, 310–11
  Mi-4P, 161–7, 310
  Mi-4S, 161–2, 310

Mi-6, 30, 34–6, 168–77, 183, 279, 295
Mi-6P, 26, 169, 171, 175–7
Mi-8, 26, 35–6, 167, 178–83, 277
Mi-10, 30, 34–6, 40, 168, 183–94
Mi-10K, 168, 189, 191–4, 295
Misrair, 80, 82, 86, 89, 290, 299
Mongolian Airlines, 43, 81, 89, 106, 162, 290, 299, 303
Morava, see L-200D
Morskaya Aviatsiya, 90, 92
Moskva, see Ilyushin Il-18 (turbines)
Moskvich, 152–3
MP-1, 89
MP-7, 89

Pan-African Air Services, 81
Pchelka, see Antonov An-14
Petroleum Helicopters, 40, 182, 190
Po-2, 14–15, 20, 22–3, 89, 278
Polskie Linie Lotnicze (LOT), 81, 84, 89, 100, 104, 107, 116, 119, 123, 125, 135, 235, 290–2, 299, 301, 305, 309
Polskie Zakłady Lotnicze (PZL), 153, 158
Polyarnaya Aviatsiya, 22, 61
PS-9, see ANT-9
PS-30, 89
PS-35, see ANT-35
PS-84, 96
PS-124, 215
PZL-101 Gawron, 256

Records, 36, 91, 152, 158, 168, 178–9, 183, 190, 204, 266, 268, 293–5
Rohrbach Roland, 21
Rossiya, see Tupolev Tu-114
Royal Nepal Airlines, 43, 162, 167, 297, 311

Schavrov
  Sch-2, 15, 89
  Sch-7, 89
Sikorsky *Il'ya Muromets*, 20, 215
SKh-1, 41
SM-1, 152
SM-1W, 152
SM-1WS, 152
SM-1WSZ, 152
SM-1WZ, 152
SM-2, 152
Stal-3, 13
Supersonic transport, see Tupolev Tu-144
Syrian Arab Airlines, 81

Tabso, 43, 69, 81, 89, 103, 105, 107, 116, 124–5, 135, 154, 290–2, 299, 305, 309

Tarom, 81, 107, 116, 291–2, 300, 305, 309
Transaviatsia, 21
Tupolev
  Tu-4, 195
  Tu-16, 198, 202
  Tu-20, 216
  Tu-70, 33, 195–7
  Tu-75, 197
  Tu-104, 24–6, 30, 32, 53, 73, 78, 113, 126, 196, 198–214, 216, 225, 231, 246, 269, 278, 288, 311–13
  Tu-104A, 25, 51, 115–16, 200–14, 225, 277–8, 280–1, 287, 292, 311–13
  Tu-104B, 25, 30, 163, 204, 207–9, 211–15, 225, 246, 277–81, 287–8, 312–13
  Tu-104D, 208, 212, 277, 279, 288, 313
  Tu-104E, 208
  Tu-104G, 199
  Tu-110, 51, 113, 128, 207, 212–15
  Tu-114, 25, 30–2, 129, 134, 215–24, 278, 280–1, 288, 292–4, 313
  Tu-114D, 216, 222–4
  Tu-124, 25, 30, 32, 37, 61, 225–35, 237–8, 246, 269, 277–80, 288–9, 292, 313–14
  Tu-124A, 233
  Tu-124K, 227, 232, 279
  Tu-124V, 226–33, 277–9
  Tu-134, 26, 30, 85, 232–40, 246, 278, 281, 314
  Tu-134A, 239
  Tu-144, 26, 35, 240–5, 248
  Tu-154, 26, 58, 246–51
  Tu-164, 240

U-2, *see* Po-2
Ukamps, 107
Ukraina, *see* Antonov An-10
Ukrvozdukhput, 21
United Arab Airlines, 107, 116, 299, 305
USSR
  Administrative areas, 12–13
  Agriculture, 18–19
  Area, 11–12
  Climate, 15, 18
  Distances, 26
  Economic regions, 13
  Fares, 26
  Industry, 18–19
  Journey times, 23, 26, 99, 210
  Population, 14
  Power, 19
  Terrain, 15
  Transport, surface, 19–20

V-2, *see* Mil Mi-2
V-8, *see* Mil Mi-8
V-10, *see* Mil Mi-10
VEB Il-14P, 103, 106–7, 304–5
Vickers Vimy Commercial, 21

Wytwornia Sprzetu Komunikacyjnego, 49, 152

Yakovlev
  Yak-12, 22, 28, 30–1, 45, 65, 152, 252–6, 269, 279, 314
  Yak-12A, 253–6, 314
  Yak-12M, 253–6, 314
  Yak-12R, 252, 256, 314
  Yak-16, 257–8
  Yak-18, 259–61
  Yak-18A, 259
  Yak-18P, 259
  Yak-18PM, 259
  Yak-18T, 259–61
  Yak-18U, 259
  Yak-24, 160, 262–9
  Yak-24A, 263–4, 268–9
  Yak-24K, 265, 268
  Yak-24P, 266–9
  Yak-24U, 263, 268–9
  Yak-40, 26, 95, 269–75
  Yak-100, 262
Yemen Airlines, 107, 154, 162, 305

Zakavia, 21